IEE CONTROL ENGINEERING SERIES 46

Series Editors: Professor P. J. Antsaklis
Professor D. P. Atherton
Professor K. Warwick

Neural networks for control and systems

Other volumes in this series:

Volume 1 **Multivariable control theory** J. M. Layton
Volume 2 **Elevator traffic analysis, design and control** G. C. Barney and S. M. dos Santos
Volume 3 **Transducers in digital systems** G. A. Woolvet
Volume 4 **Supervisory remote control systems** R. E. Young
Volume 5 **Structure of interconnected systems** H. Nicholson
Volume 6 **Power system control** M. J. H. Sterling
Volume 7 **Feedback and multivariable systems** D. H. Owens
Volume 8 **A history of control engineering, 1800-1930** S. Bennett
Volume 9 **Modern approaches to control system design** N. Munro (Editor)
Volume 10 **Control of time delay systems** J. E. Marshall
Volume 11 **Biological systems, modelling and control** D. A. Linkens
Volume 12 **Modelling of dynamical systems—1** H. Nicholson (Editor)
Volume 13 **Modelling of dynamical systems—2** H. Nicholson (Editor)
Volume 14 **Optimal relay and saturating control system synthesis** E. P. Ryan
Volume 15 **Self-tuning and adaptive control: theory and application** C. J. Harris and S. A. Billings (Editors)
Volume 16 **Systems modelling and optimisation** P. Nash
Volume 17 **Control in hazardous environments** R. E. Young
Volume 18 **Applied control theory** J. R. Leigh
Volume 19 **Stepping motors: a guide to modern theory and practice** P. P. Acarnley
Volume 20 **Design of modern control systems** D. J. Bell, P. A. Cook and N. Munro (Editors)
Volume 21 **Computer control of industrial processes** S. Bennett and D. A. Linkens (Editors)
Volume 22 **Digital signal processing** N. B. Jones (Editor)
Volume 23 **Robotic technology** A. Pugh (Editor)
Volume 24 **Real-time computer control** S. Bennett and D. A. Linkens (Editors)
Volume 25 **Nonlinear system design** S. A. Billings, J. O. Gray and D. H. Owens (Editors)
Volume 26 **Measurement and instrumentation for control** M. G. Mylroi and G. Calvert (Editors)
Volume 27 **Process dynamics estimation and control** A. Johnson
Volume 28 **Robots and automated manufacture** J. Billingsley (Editor)
Volume 29 **Industrial digital control systems** K. Warwick and D. Rees (Editors)
Volume 30 **Electromagnetic suspension—dynamics and control** P. K. Sinha
Volume 31 **Modelling and control of fermentation processes** J. R. Leigh (Editor)
Volume 32 **Multivariable control for industrial applications** J. O'Reilly (Editor)
Volume 33 **Temperature measurement and control** J. R. Leigh
Volume 34 **Singular perturbation methodology in control systems** D. S. Naidu
Volume 35 **Implementation of self-tuning controllers** K. Warwick (Editor)
Volume 36 **Robot control** K. Warwick and A. Pugh (Editors)
Volume 37 **Industrial digital control systems (revised edition)** K. Warwick and D. Rees (Editors)
Volume 38 **Parallel processing in control** P. J. Fleming (Editor)
Volume 39 **Continuous time controller design** R. Balasubramanian
Volume 40 **Deterministic control of uncertain systems** A. S. I. Zinober (Editor)
Volume 41 **Computer control of real-time processes** S. Bennett and G. S. Virk (Editors)
Volume 42 **Digital signal processing: principles, devices and applications** N. B. Jones and J. D. McK. Watson (Editors)
Volume 43 **Trends in information technology** D. A. Linkens and R. I. Nicolson (Editors)
Volume 44 **Knowledge-based systems for industrial control** J. McGhee, M. J. Grimble, A. Mowforth (Editors)
Volume 45 **Control theory: a guided tour** J. R. Leigh

Neural networks for control and systems

Edited by
K. Warwick, G. W. Irwin
and K. J. Hunt

Peter Peregrinus Ltd. on behalf of the Institution of Electrical Engineers

Published by: Peter Peregrinus Ltd., London, United Kingdom

Peter Peregrinus Ltd.,
Michael Faraday House,
Six Hills Way, Stevenage,
Herts. SG1 2AY, United Kingdom

British Library Cataloguing in Publication Data

A CIP catalogue record for this book
is available from the British Library

ISBN 0 86341 279 3

Printed in England by Short Run Press Ltd., Exeter

Contents

List of contributors

G. Lightbody
G.W. Irwin
Department of Electrical and Electronic Engineering
Queen's University of Belfast

C. Hall
R. Smith
Cambridge Consultants Ltd.
Cambridge

K.J. Hunt
P.J. Gawthrop
D. Sbarbaro
R. Zbikowski
Department of Mechanical Engineering
University of Glasgow

A. Redgers
I. Aleksander
Department of Electrical Engineering
Imperial College

Q.H. Wu
Department of Mathematical Sciences
Loughborough University of Technology

N. Dodd
Neural Solutions
Cheltenham

G.A. Montague
A.J. Morris
M.J. Willis
Department of Chemical and Process Engineering
University of Newcastle upon Tyne

K. Warwick
R.J. Mitchell
J.M. Bishop
Department of Cybernetics
University of Reading

S. Billings
S. Chen
Department of Automatic Control and Systems Engineering
University of Sheffield

J.C. Mason
P.C. Parks
Royal Military College of Science
Shrivenham

C.J. Harris
Department of Aeronautics and Astronautics
University of Southampton

Preface

Neural networks for control and systems has blossomed as a topic over the last few years, not only in terms of the research being carried out, but perhaps more importantly in terms of the potential range of applications. The aim of this book is to present an introduction to, and an overview of, the present state of neural network research and development with particular reference to systems and control application studies, in a fairly broad context covering a wide variety of network types and implementation areas. The form of the book is such that it should be useful for various levels of reader, with earlier chapters covering basic principles and fundamental design procedures, through more advanced, detailed structures with finally a look at application possibilities.

The book arises out of an IEE Workshop on Neural Networks for Control and Systems : Principles and Applications, at the University of Reading, April 1992, being directed as a book to support the material presented at the Workshop. In fact the Workshop actually came about as a result of a series of IEE meetings in this subject area. The first event was a colloquium held at Savoy Place, London, which attracted a record number of participants. The second was a two day IEE Workshop held in Prague, Czechoslovakia, the first ever event staged by the IEE in Eastern Europe, this also being a considerable success, attracting over 120 delegates.

The level of the book makes it suitable for practising engineers, final year degree students and both new and advanced researchers in the field. The book could easily be a core text for Masters or final year programmes covering artificial neural networks, and their implementation in their broadest context. The contents of the book have been selected so as to reduce the amount of abstract theoretical ideas and to provide sensible suggestions for implementation and application areas.

The earlier chapters include a good introductory overview, Chapter 1 in particular being a gentle guide to the field for those not necessarily familiar with the subject area. Chapter 2, however, takes a look at n- tuple digital networks in particular, these being encountered chiefly in the form of a hardware base. A survey of neural networks directed towards implementation is then given in Chapter 3, thereby indicating the broad spectrum of application possibilities.

Chapters 4 to 8 concentrate on different aspects of neural networks applied within a control systems environment. In Chapter 4 feedforward networks are considered in terms of adaptive control. This chapter also includes an interesting case study in the form of turbo- generator control. Chapter 5, meanwhile, looks at real-time control and compares neural networks with fuzzy logic, concluding with the description of neural networks applied to autonomous land vehicle control. The following chapter links neural networks more closely with control procedures such as system identification, learning algorithms and Gaussian structures. Again the chapter concludes with an example, this time pH control. In Chapter 7 various neural network applications in the area of process control are described, explaining radial basis functions, statistical methods and giving a comparison between static and dynamic networks.

The two chapters which follow, 8 and 9, give more background on the theory of neural networks by looking at advanced techniques. Chapter 8 is comprehensive in its coverage of approximation theory, looking at network structures and algorithm convergence. Chapter 9 then looks more deeply into learning algorithms, radial basis functions and system identification, thereby providing much rigour to the earlier work.

The final three chapters are very much concerned with application studies. In Chapter 10 a variety of applications is presented, including vision systems, colour recipe prediction and speech processing. Chapter 11 then discusses the choice of network structure for applications, and considers the advantages in the selection of an optimal network. The last Chapter considers the different pitfalls in the use and selection of neural networks and asks the pertinent question, 'why use neural networks?'

As a whole, the contributions which make up this book provide a good coverage of the present state of neural network research and implementation. It is possible for the book to be used as a complete text, comprehensively covering the field, or in terms of specific chapters detailing areas of particular interest. It is felt that the book is especially useful, however, for those with a systems and/or control background who wish to find out what neural networks are, and whether or not such networks could be use of use to them.

The Editors would like to thank all of the authors for their contributions and the very prompt attention that each of them gave to the production of their finalised text. They would also like to thank those at Peter Peregrinus Limited who were responsible for the book appearing in good time, along with advice from the Series Editors, Derek Atherton and Panos Antsaklis. Finally the Editors would like to express their gratitude to Angela Stanfield and Sally Drewe from the IEE for their organisational help at the Workshop and to Liz Lucas for her help in putting the text together.

Kevin Warwick
George Irwin
Kenneth Hunt

February 1992.

Chapter 1

Neural networks: an introduction

K. Warwick

1.1 Introduction

The study of neural networks is an attempt to understand the functionality of the brain. In particular it is of interest to define an alternative 'artificial' computational form that attempts to mimic the brain's operation in one or a number of ways, and it is this area of artificial neural networks which is principally addressed here.

Essentially artificial neural networks is a 'bottom up' approach to artificial intelligence, in that a network of processing elements is designed, these elements being based on the physiology and individual processing elements of the human brain. Further, mathematical algorithms carry out information processing for problems whose solutions require knowledge which is difficult to describe.

In the last few years interest in the field of neural networks has increased considerably, due partly to a number of significant break-throughs in research on network types and operational characteristics, but also because of some distinct advances in the power of computer hardware which is readily available for net implementation. It is worth adding that much of the recent drive has, however, arisen because of numerous successes achieved in demonstrating the ability of neural networks to deliver simple and powerful problem solutions, particularly in the fields of learning and pattern recognition, both of which have proved to be difficult areas for conventional computing.

In themselves, digital computers provide a media for well defined, numerical algorithm processing in a high performance environment. This is in direct contrast to many of the properties exhibited by biological neural systems, such as creativity, generalisation and understanding. However computer-based neural networks, both of hardware and software forms, at the present time provide a considerable move forward from digital computing in the direction of biological systems, indeed several biological neural system properties can be found in certain neural network types. This move is supported by a number of novel practical examples, even though these tend to be in fairly scientific areas, e.g. communication processing and pattern recognition.

Because of its inter-disciplinary basis, encompassing computing, electronics, biology, neuropsychology etc, the field of neural networks attracts a variety of interested researchers and implementers from a broad range of backgrounds. This makes the field very exciting with a flood of new ideas still to be tried and tested. In this chapter a brief view is given of some of the various artificial neural network techniques presently under consideration, a few application areas are discussed and indications are given as to possible future trends.

1.2 Neural network principles

The present field of Neural Networks links a number of closely related areas, such as parallel distributed processing, connectionism and neural computing, these being brought together with the common theme of attempting to exhibit the method of computing which is witnessed in the study of biological neural systems. An interesting and succinct definition of neural computing (Aleksander and Morton, 1990) is that it is a study of networks of adaptable nodes which, through a process of learning from task examples, store experiential knowledge and make it available for use. Taking this definition alone, although representing a significant step forward from standard digital computing, it does indicate a rather limited, practically oriented state at the present time, leaving much ground still to be covered towards the artificial reconstruction of complex biological systems.

A fundamental aspect of artificial neural networks is the use of simple processing elements which are essentially models of neurons in the brain.

These elements are then connected together in a well structured fashion, although the strength and nature of each of the connecting links dictates the overall operational characteristics for the total network. By selecting and modifying the link strengths in an adaptive fashion, so the basis of network learning is formed along the lines of the previous (Aleksander and Morton, 1990) definition.

One important property of a neural network is its potential to infer and induce from what might be incomplete or non-specific information. This is however also coupled with an improvement in performance due to the network learning appropriate modes of behaviour in response to problems presented, particularly where real-world data is concerned. The network can, therefore, be taught particular patterns of data presented, such that it can subsequently not only recognise such patterns when they occur again, but also recognise similar patterns by generalisation. It is these special features of learning and generalisation that make neural networks distinctly different from conventional algorithm processing computers, along with their potential property of faster operational speeds realised through inherent parallel operation.

1.3 Neural network elements

It is not the intention here to provide a detailed history of artificial neural networks, however earlier stages of development can be found in Hebb (1949), Rosenblatt (1962) and Minsky and Papert (1969), whereas a good guide to more recent events can be picked up in Rummelhart and McClelland (1986) and Pao (1989). The discussion here focuses on some of the different approaches taken in the design of neural networks and their basic principles of operation.

In general, neural networks consist of a number of simple node elements, which are connected together to form either a single layer or multiple layers. The relative strengths of the input connections and also of the connections between layers are then decided as the network learns its specific task(s). This learning procedure can make use of an individual data set, after which the strengths are fixed, or learning can continue throughout the network's lifetime, possibly in terms of a faster rate in the initial period with a limited amount of 'open mind' allowance in the steady- state. In certain networks a supervisor is used to direct the learning procedure, and this method is referred to as 'supervised learning', whereas it is also possible to realise self-organising networks which learn in an unsupervised fashion.

The basic node elements employed in neural networks differ in terms of the type of network considered. However one commonly encounted model is a form of the McCulloch and Pitts neuron (Aleksander, 1991), and is shown in Figure 1.1. In this the inputs to the node take the form of data items either from the real-world or from other network elements, possibly from the outputs of nodes in a previous layer. The output of the node element is found as a function of the summed weighted strength inputs.

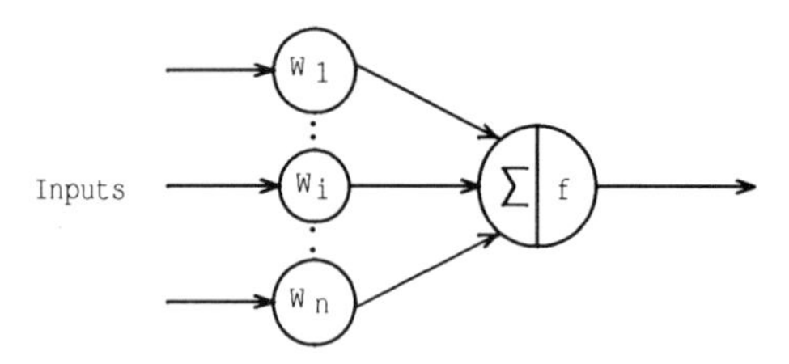

Figure 1.1 Basic Neuron Model

This output signal can then be employed directly, or it can be further processed by an appropriate thresholding or filtering action, a popular form being a sigmoid function. For some networks the node output signal is therefore used directly, whereas in other cases the important feature is simply whether or not the node has fired due to the weighted sum exceeding a previously defined threshold value for that particular node.

1.4 Hopfield networks

In the network model proposed by Hopfield (1982), binary input signals are introduced such that each of the network node elements has an output which acquires one of two possible states, either +1 or -1. Further, the number of node elements is assumed to be extremely large, with each element connected, in some way, to all others, i.e. the output of a node is connected as a weighted input to all other nodes. Each node does, however, have its own separate input which is unique, and this is an effective input interface with the real-world.

A signal/data pattern can be applied to the real world network inputs, and the weights adjusted, by such means as Hebbian learning, to produce an output pattern which is, in some way, associated with that particular input pattern. A good description, in slightly more detail, of this operation can be found in Benes (1990), in which it is shown how weights are adjusted to arrive at a local minimum of a defined global energy function. This energy function is given by

$$E = -\frac{1}{2} \sum_{i,j=1}^{N} b_{ij} S_i S_j \qquad (1.1)$$

where E is the energy function, N is the total number of nodes, S_i and S_j are the output values, either +1 or -1, of the ith and jth node elements and b_{ij} is the weighting applied to the link from the jth node output to the ith node input.

1.5 Kohonen networks

In many ways Kohonen (1984) followed on directly from Hopfield by using a single layer network as a basis for speech recognition. Here a speech signal, in the form of a phoneme, is considered with regard to the energy content of the phoneme at different frequencies. In fact the phoneme energy is split up into several frequency components by means of bandpass filters, the output of each filter acting as an input to the net. By this means, when a phoneme is uttered by a particular person, a certain pattern is presented as an overall input to the network.

For a pattern presented as input to the network, a clustering effect on the network output can be enforced by adjustment of the weightings between nodes. This can be achieved by selecting a node output as that which resembles a specific phoneme. The nodes which are adjacent then have their weights adjusted, the amount of adjustment being dependent on their distance from the selected node. If the selected node weighting is adjusted by an amount x, then the adjacent nodes' weightings are adjusted by an amount Q, where Q is given by the Mexican hat function shown in

Figure 1.2. In this way a tailored network response, to a presented input pattern, is enforced, i.e. the network is taught how to respond to a certain input pattern.

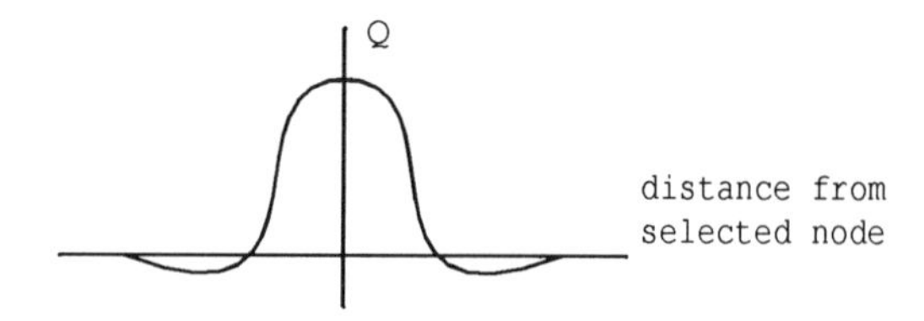

Figure 1.2 Mexican Hat

In the Kohonen network, the nodes can be laid out in a two dimensional matrix pattern, such that when a phoneme pattern is input to the network, a small area of the matrix, centred around one node output, should exhibit active outputs; the actual area being selected due to the training/tailoring process. In this way, having been trained on numerous phoneme energy inputs, when a phoneme is uttered again, this should be recognisable on the matrix. Problems arise due to (a) different speakers, therefore different energy spectra, of the same phoneme, (b) the same speaker uttering a phoneme in a different way and (c) phonemes that are new to the matrix. It is however possible (Wu and Warwick, 1991) to form a hierarchical system where prior knowledge is available of the likely phoneme to be uttered, such cases arising in natural language.

1.6 Back propagation networks

Back propagation networks have very quickly become the most widely encountered artificial neural networks, particularly within the area of systems and control. In these networks it is assumed that a number of node element layers exist and further, that no connections exist between the node elements in a particular layer. One layer of nodes then forms the input layer whilst a second forms the output layer, with a number of intermediate or hidden layers existing between them. It is often the case, however, that only one, two or even no hidden layers are employed.

A back propagation network requires a pattern, or set of data, to be presented as inputs to the input node element layer. The outputs from this layer are then fed, as weighted inputs, to the first hidden layer, and subsequently the outputs from the first layer are fed, as weighted inputs, to the second hidden layer. This construction process continues until the output layer is reached. The structure is shown in simple terms in Figure 1.3, although it should be stressed that this arrangement only includes one hidden layer and further, no input weightings are indicated.

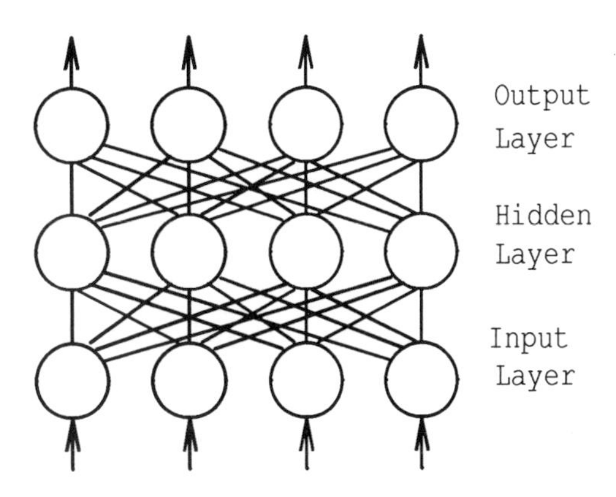

Figure 1.3 Back Propagation Net (feedforward part)

In some ways, training of a back propagation network is similar to that for a Kohonen network, in that for a given input pattern the network is forced to take on a particular output form by means of weighting selection. Each output layer node value can, for example, be compared with a desired or specified set value, the difference between the two signals being realised as an error. As part of a vector of, similar type, errors this signal can then be minimised, possibly in a least squares sense, by adjustment of the node element input weightings for each of the layers. Numerous algorithms then exist, by means of which the weightings can be appropriately adjusted in response to the error values. It is worth noting however, that due to important factors such as noise and numerical problems, training of the network is not a simple task.

Such networks can be made use of in a number of ways in the systems and control area (Willis et al, 1991) one example being in the formation of process models, where reasonably cost effective and reliable results can be achieved. The back propagation network is especially useful in this respect, because of its inherent non-linear mapping capabilities, which can deal effectively with a wide range of process features. Another immediate use of such networks is in the real-time estimation of process parameters within an adaptive model of the plant under control.

The employment of neural network models of a process plant can offer distinct advantages when compared with more conventional techniques. As an example, networks can be used to obtain frequent measurement information of a form which is better than that which can be obtained by hardware methods alone, thus overcoming some problems incurred with delayed signals. Such networks can also be used to provide inferred values for signals which are difficult to measure in a practical situation, something which is a common occurrence in many process control problems

The use of artificial networks within model based control schemes has also been considered, such as those employing a model reference strategy (Narendra and Parthasarathy, 1990), a useful mode of implementation being in the form of a multi-layered network processor, something which is particularly beneficial when applied to nonlinear systems.

Back propagation networks are also well suited for real-time on-line computer control systems because of their abilities in terms of rapid processing of collected plant input-output data. One example of this is in the control of a robot manipulator (Warwick, 1991), where the network mapping adapts, in terms of its weightings, as it learns with regard to characteristic changes in robot manipulator joints, such changes occurring as the manipulator moves within its field of operation. The network role is to produce an output vector which includes signals such as servo motor input power and desired trajectories. Network weightings are adapted on-line as manipulator operating conditions change, such that once an initial learning phase has been completed the network is allowed to continue an element of learning as it is in operation, in order to track changes in motor joint characteristics. This is shown schematically in Figure 1.4.

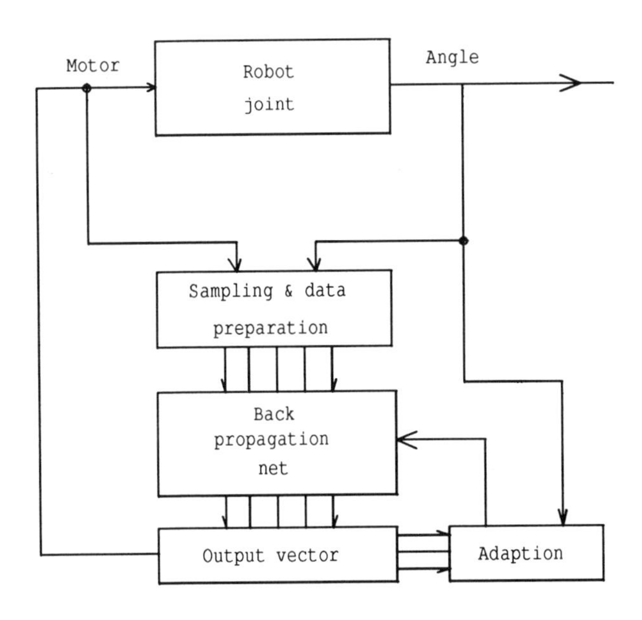

Figure 1.4 Neural net for robot joint control

1.7 N-tuple networks

N-tuple networks, also known as digital nets, are based on a slightly different neuron model and also a different learning strategy to those methods already discussed (Aleksander (1989) and (1991)). Each node element makes use of binary data and the node is actually designed in terms of digital memory. Essentially the node inputs are memory address lines whilst the output is the value of the data stored at that address, as shown in Figure 1.5.

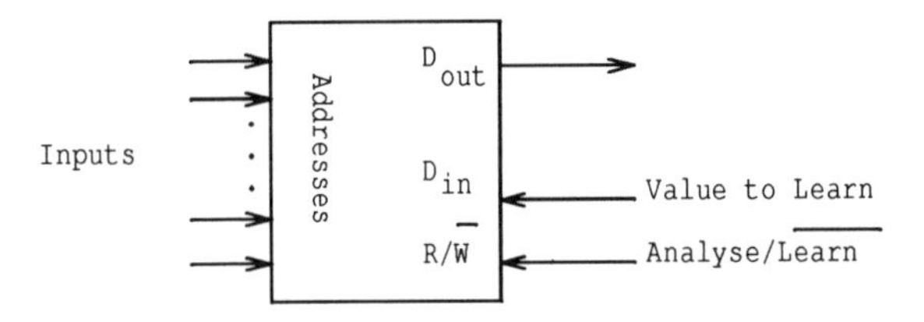

Figure 1.5 RAM Neuron

When the node element is in its learning mode, the pattern/data being taught is entered on the memory address lines, and the appropriate value stored. When in analysis mode, the node is addressed and the stored value thus appears at the node output.

Figure 1.6 shows schematically how input patterns are fed to the n-tuple neural network, such that for each node element, n bits of the input pattern are sampled (as an n-tuple). The first n-tuple is used to address the first node element, such that a '1' is stored at that address. A further sampled n-bits are learnt by the second node element, and so on until the whole input pattern has been dealt with in a random fashion.

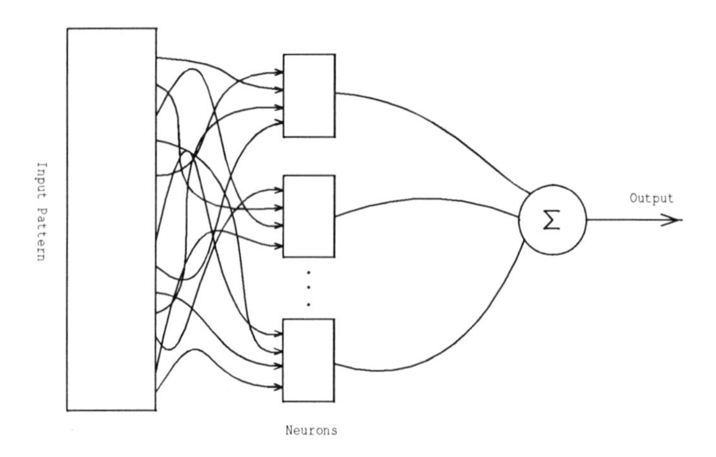

Figure 1.6 n-tuple network

When the n-tuple network is required to analyse a pattern, once a particular input pattern has been presented, the number of node elements outputting a '1' is counted. This means that if an 'exactly' identical input pattern is presented, when compared with the learnt pattern, the count will be 100%, whereas if the pattern is similar the count will be very high. In practice the neural network can be considered to 'recognise' an object if the count is greater than x%, where x is some previously defined confidence value. If a new input pattern needs to be learnt however, the network can be cleared by setting all node elements to zero and starting the procedure again.

1.8 Conclusions

To regard the presently available neural net techniques as being a solution to the search for true artificial intelligence would be a falsehood, but rather they are techniques which exhibit certain intelligent properties and are useful for solving some specific tasks. Indeed many 'intelligent' systems combine features of other artificial intelligent tools such as expert systems or genetic algorithms along with neural nets.

An overall intelligent system, however, requires much more than a central processing unit, which is the part played by a neural net, such that sensory systems, genetic features and problem understanding are required along with their appropriate interface to the central unit.

At present neural nets are extremely powerful, rapidly operating devices which exhibit many advantages when compared with more conventional computing methods. They can be constructed in a number of forms, some of which have been described here, and are already in use in a wide variety of applications.

The number of possible neuron models is however very large with more complex models representing the operation of human neurons more closely. The artificial neural networks considered here, and those which are generally in use for systems and control, tend to be much simpler and easier to realise physically.

1.9 References

Aleksander, I., (ed.), 1989, 'Neural computing architectures', North Oxford Academic Publishers.

Aleksander, I., 1991, 'Introduction to neural nets', Chapter 9 in 'Applied Artificial Intelligence', K. Warwick (ed.), Peter Peregrinus Ltd.,

Aleksander, I., and Morton, H., 1990, 'An introduction to neural computing', Chapman and Hall.

Benes, J., 1990, 'On neural networks', Kybernetica, **26**, No. 3, pp. 232-247.

Hebb, D.E., 1949, 'Organisation of behaviour', Wiley.

Hopfield, J.T., 1982, 'Neural networks and physical systems with emergent collective computational abilities', Proc. Natl. Acad. Sci. USA 79, pp. 2554-2558.

Kohonen, T., 1984, 'Self-organisation and associative memory', Springer-Verlag.

Minsky, M., and Papert, S., 1969, 'Perceptrons', MIT Press.

Narendra, K.S., and Parthasarathy, K., 1990, 'Identification and control of dynamical systems using neural networks', IEEE Trans. on Neural Networks, **NN-1**, 1, pp. 4-27.

Pao, Y-H, 1989, 'Adaptive pattern recognition and neural networks', Addison Wesley.

Rosenblatt, F., 1962, 'Principles of neurodynamics', Spartan.

Rumelhart, D.E., and McClelland, J.L., 1986, 'Parallel distributed processing', MIT Press, Vols. 1 and 2.

Warwick, K., 1991, 'Neural net system for adaptive robot control', in 'Expert Systems and Robotics', T. Jordanides and B. Torbey (eds.), Springer- Verlag, Computer Systems Science Series, **71**, pp. 601-608.

Willis, M.T., Di Massimo, C., Montague, G.A., Tham, M.T., and Morris, A.J., 1991, 'Artificial neural networks in process engineeering', Proc. IEE, Part D, **138**, 3, pp. 256-266.

Wu, P., and Warwick, K., 1991, 'A structure for neural networks with artificial intelligence', in 'Research and Development in Expert Systems VIII', I.M. Graham and R.W. Milne (eds.), Cambridge University Press, pp. 104-111.

Chapter 2

Digital neural networks

A. Redgers and I. Aleksander

2.1 Classification of artificial neural networks

2.1.1 Introduction: McCulloch and Pitts versus boolean nodes

When McCulloch and Pitts first studied artificial neural networks, **ANNs**, their neuron model consisted of binary signals contributing to a sum which was then thresholded to produce the output of the neuron. This model quickly evolved to the well known "function of weighted sum of inputs" model, the function usually being a sigmoidal squashing function such as $output = \frac{1}{1+\exp(-\lambda(\text{sum}-\text{bias}))}$ or $output = \text{erf}(\lambda(\text{sum}-\text{bias}))$.

The binary threshold function $output = \Theta(\text{sum}, \text{bias})$ can be thought of as the limiting case of these when the parameter λ becomes large. Function-of-weighted-sum-of-inputs neuron models are collectively termed McCulloch and Pitts, **McCP**, models to distinguish them from other models. The word **node** is equivalent to the phrase "artificial neuron."

The weights and bias can be considered to be other parameters of McCP nodes which can be set, **trained**, so that a network of McCP nodes performs the desired function of its inputs. However, there are functions of their inputs which McCP nodes cannot perform. For example, there is no setting of weights and threshold (bias) which will allow a 2-input McCP neuron to perform the XOR function: "fire if and only if a single input is set to one."

Boolean nodes are based on random access memories, **RAMs**, which are look-up tables. The inputs form the address of a memory location and the RAM outputs the contents of that location during operation or changes it during training. RAMs can perform any binary function of their inputs, but they do not scale well as the number of addressable memory locations increases as 2^{inputs}. In practice they are **decomposed**, split into **Large Logic Neurons, LLNs**, consisting of smaller boolean nodes.

Conventionally the contents of memory locations in RAMs are binary valued, recent research has focussed on boolean models with three or more valued contents. The output of the node is then a *function* of the contents of the addressed memory location. In the case of RAMs this function is just the identity function; in the case of three-valued **Probabilistic Logic Nodes, PLNs**, the function outputs 1 and 0 for memory contents 1 and 0, but outputs 1 or 0 randomly for memory contents U, the third value. **Multi-PLNs, MPLNs**, have even more allowed values, **activation states**, of memory contents.

2.1.2 Definition of ANNs

The definition "an interconnected system of parameterised functions" covers many types of ANNs and neuron models. For example, in McCP and boolean nodes the parameters are respectively: the weights (and bias), and the memory contents. The functions are respectively, "output a function of the weighted sum of inputs," and "output a function of the contents of the addressed memory location."

It is useful to be able to classify ANNs in order to compare them and place them relative to each other in our understanding. Attempts have been made to do this such as that in Rumelhart, Hinton & McClelland (1986). One simple classification scheme which achieves this requires five pieces of information for the description of an ANN as defined above:

- **Topology** - how the functions are interconnected.
- **Architecture** - use / type of network (four possibilities).
- **Neuron model** - what the functions are.
- **Training algorithm** - how the parameters are set.
- **Operation schedule** - timing of function interactions.

The topology of an ANN is a set of connections between inputs, outputs and nodes. All topologies are of course subsets of the fully connected topology - any topology can be considered as just the fully connected topology with the node functions independent of certain inputs. Architecture is more vague - it is the user's intention of how the ANN is to work. Simple architectures come in four main types: **feedforward, external feedback, internal feedback** and **unsupervised** (figure 2.1).

With external feedback the previous output provides the new input. There is only one external stimulus, at the very start, and any new stimulus elicits the same response no matter what other stimuli have previously been shown to the network.

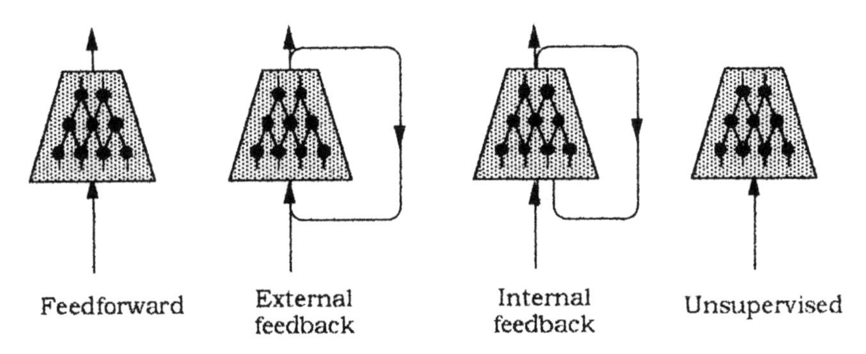

figure 2.1 - four possible architectures of ANNs

With internal feedback the net sees both an external stimulus and its own response to the previous stimulus. Its response depends not only on the current stimulus but also on the previous stimuli and the order in which they were presented to the net. More complex architectures are formed of these simple components. It may be that there are still other architectures, that cannot be classified using any of the four existing possibilities, which should be added to this list.

The McCulloch and Pitts and boolean models already mentioned are examples of frequently studied neuron models. The training algorithm conducts a search through the space of parameter values for a set of values with which the network will perform the required function. Networks of McCP nodes are frequently trained using variations of Rumelhart's Error Back Propagation. Several algorithms have been developed for supervised training of a tree, **pyramid** (see figure 2.2), network of PLNs, e.g. Myers (1988), Al-Alawi (1989b), and Filho, Bisset and Fairhurst (1990). All have similar characteristics, Myers' algorithm being the simplest. It is just the following:

1 Present a new training example to the input nodes at the base of the pyramid.
tries := 1
2 Allow signals from the training example to propagate through

the network until they reach the output node at the top of the pyramid.

3 If the output is correct then

In each node if the addressed location contains a U then set it to the (random) value that the node has just output, **gel**

Else

If **tries** < maxTries then

tries := **tries** + 1

Goto step 2

Else

Set contents of addressed locations to U, **punish**

4 Goto step 1

In Al-Alawi's and Filho et al's training algorithms a flag is set for each node to say whether its output was probabilistic. If the output is incorrect the search for improving node contents is conducted on those nodes whose output was probabilistic.

The operation schedule is a list of times and neurons which respond at those times. In a layered feedforward network the neurons closest to the retina respond first, their responses feed into the neurons of the next layer, which then respond, and so on. In networks with feedback the set of neurons responding at a given time may be chosen randomly. In a **synchronous** feedback network all nodes respond at the same time. In an **asynchronous** network only one node responds at a time. But any number of nodes, **synchronicity**, between these two extremes is possible. A synchronicity of 1 means asynchronous; a synchronicity of N, the number of nodes in the network, means synchronous.

2.1.3 Limitations of the classification scheme

There may be ANNs which do not fit into this classification scheme, or whose inclusion is strained or trivial. For example in networks such as those of Binstead and Jones (1987), nodes are added during training, that is, the topology changes. But these can be regarded as fully connected networks with functions at first independent of, but later dependent on, certain inputs as extra parameters are allowed to change.

Compiled networks, such as Hopfield networks (Tank & Hopfield 1987), do not have an iterative training procedure, instead the node function parameters are calculated explicitly from the data examples. Either such off-line procedures can be considered to be training algorithms of a sort, or else one can regard the network as not having a training algorithm.

One normally expects that a function called twice with the same arguments will give the same answer both times - not so if the function has an **internal state** which changes every time the function is called. Gorse & Taylor (1988, 1990) have proposed neuron models whose functions change continuously over time, for example modelling refractory periods in brain neurons. Although these can be covered by considering internal states of nodes to be parameters of the functions there is a case for adding a sixth category to the classification scheme above.

2.2 Functions of boolean nodes

2.2.1 MPLN activation functions

PLNs are 3-valued RAMs with the third, U, activation state representing a 0.5 probability of outputting 1. MPLNs have more activation states each with their own probability of outputting 1. The contents of memory locations are changed to neighbouring activation states during training, providing an ordering on the set of activation states. Making the activation function, mapping activation states to probabilities of outputting 1, sigmoidal rather than linear reduces the search space. Myers (1989) has obtained results indicating that for MPLNs the threshold function is an efficient function of the activation state s:

$$f(s) = \begin{cases} s < U : f(s) = 0 \\ s = U : f(s) = 0.5 \\ s > U : f(s) = 1 \end{cases}$$

This looks like a PLN with extensions to either side of the U value, and during training the extensions serve to keep a count of how many times a value is used correctly.

2.2.2 The address decoder function

Martland (1989) gives the probability, r, of a 2-input PLN outputting a 1 when addressed as:

$$r = a_{00}(1-x_1)(1-x_2) + a_{01}(1-x_1)x_2 + a_{10}x_1(1-x_2) + a_{11}x_1x_2 \qquad (2.1)$$

where:

a_{00} is the contents of address location 00, i.e. the probability of the PLN firing if it is addressed with 00. Similarly for a_{01}, a_{10} and a_{11}.

x_1, x_2 are the (binary) values on the two input lines.

The inputs x_1 and x_2 need not be just binary but can take any real value between 0 and 1; in Taylor and Gorse's pRAM they represent *probabilities* of the inputs being 1. The resulting probability r might not even be run through a generator to obtain 1 or 0, but instead passed on directly. Martland himself used the address decoder formula (2.1) to implement the Error Back Propagation algorithm on pyramids of these 'continuous RAMs' (training took far longer than Myers' algorithm for 3-valued PLNs).

The address decoder function for an n-input boolean node is:

$$r = \sum_{\omega \in \mathbb{B}^n} a_\omega \prod_{i=1}^{n} v(\omega_i, x_i) \qquad (2.2)$$

where:

$v(p, q)$ is a function such that: $\begin{cases} v(1,q) = q \\ v(0,q) = 1-q \end{cases}$

e.g. $\quad v(p,q) = 2pq - p - q + 1 \qquad (2.3)$

$\mathbb{B}^n$ is the set of boolean strings of n 1's and 0's (e.g. "0101" ε $\mathbb{B}^4$).

ω is one such string (the sum is taken over all strings).

ω_i is the ith component of ω.

a_ω is the contents of the memory location whose address is ω.

x_i is the ith input to the node.

Equation (2.1) can be rewritten as a power series in x_1 and x_2:

$$r = a_{00} + x_1(a_{10}-a_{00}) + x_2(a_{01}-a_{00}) + x_1x_2(a_{11}+a_{00}-a_{10}-a_{01}) \qquad (2.4)$$

The n-input address decoder function (2.2), (2.3) can be similarly rewritten as a power series:

$$r = \sum_{\omega \in \mathbb{B}^n} g_\omega \prod_{i=1}^{n} w(\omega_i, x_i) \qquad (2.5)$$

where:

$w(p, q)$ is a function such that: $\begin{cases} w(1,q) = q \\ w(0,q) = 1 \end{cases}$

e.g. $\quad w(p,q) = pq - p + 1$ (2.6)

Given either the coefficients a_ω or g_ω one can obtain the coefficients for the other form:

$$a_\omega = \sum_{\phi \in \mathbb{B}^n} g_\phi \prod_{i=1}^{n} y(\omega_i, \phi_i) \tag{2.7}$$

where:

$$y(p,q) = pq - q + 1 \tag{2.8}$$

In the other direction:

$$g_\omega = \sum_{\phi \in \mathbb{B}^n} a_\phi \prod_{i=1}^{n} z(\omega_i, \phi_i) \tag{2.9}$$

$$z(p,q) = pq - q + 1 \tag{2.10}$$

In a McCP node only the constant term $g_{000...}$ and the first degree coefficients $g_{100...}$, $g_{010...}$, $g_{001...}$, the *weights*, are non-zero. In McCP nodes with **Π- units** (Williams, 1986) second - $g_{110...}$, $g_{101...}$, $g_{011...}$ - or higher degree coefficients are non-zero. To discover whether the function performed by a boolean node can be performed by an McCP node, transform the a_ω into g_ω and check to see which coefficients are non-zero. The reverse transformation will implement a function performed by a McCP node on a boolean node.

2.2.3 Kernel functions and transforming output functions

Equations (2.3), (2.6), (2.8) and (2.10) are examples of **kernel functions** of the form:

$$v(p, q) = hpq + kp + lq + m \tag{2.11}$$

Given any kernel function (2.11) for an output function:

$$r = \sum_{\omega \in \mathbb{B}^n} h_\omega \prod_{i=1}^{n} v(\omega_i, x_i) \tag{2.12}$$

we can obtain the kernel function for the transformation to the address decoder function (2.2), (2.3):

$$a_\omega = \sum_{\phi \in \mathbb{B}0n} h_\phi \prod_{j=1}^{n} u(\omega_j, \phi_j) \quad (2.13)$$

where:

$$u(p,q) = hpq + lp + kq + m \quad (2.14)$$

The kernel function for the transformation to (2.5), (2.6) is:

$$u(p,q) = (h\text{-}k)pq + (1\text{-}m)p + kq + m \quad (2.15)$$

The kernel function for the identity transformation from h_ω to h_ω is:

$$h_\omega = \sum_{\phi \in \mathbb{B}n} h_\phi \prod_{j=1}^{n} i(\omega_j, \phi_j) \quad (2.16)$$

is

$$i(p,q) = 2pq - p - q + 1 \quad (2.17) = (2.3)$$

If linear transformations are made of the variables x_i : $y_i = ax_i + b$, $x_i = (y_i - b)/a$ (e.g. binary $x_i \leftrightarrow$ bipolar y_i : $y_i = 2x_i - 1$, $x_i = (y_i + 1)/2$) then in the kernel function (2.11) the variable q, representing x_i, is transformed in the same way: $q := (q - b)/a$.

2.3 Large Logic Neurons

2.3.1 Decomposition of boolean nodes

The number of weights in McCP nodes increases as the number of inputs to the node; with current computers it is possible to simulate or build McCP nodes with thousands or even millions of inputs. However, the number of memory locations in boolean nodes (and weights in Π- unit models) grows exponentially with the number of inputs - to implement a node with 240 inputs would require 2^{240} bits of artificial storage which is of the order of the number of particles in the universe. Hence when scaling up boolean nodes it is necessary to **decompose** them into devices called **Large Logic Neurons, LLNs**, consisting of many small boolean nodes linked up so as to produce a single output.

There are two main types of LLN's: **discriminators**, in which the large boolean node is replaced by a single layer of small nodes whose outputs are summed (the output of the LLN is a function of that sum); and **pyramids**, in which the single large boolean node is replaced by small nodes arranged in a tree-like structure feeding finally into single node with a single output.

When nodes are decomposed in this way they are no longer **universal** i.e they cannot perform every possible (binary valued) function of their (binary) inputs. As part of an attempt to find *which* functions they can perform we have begun by calculating *how many* functions they can perform, **functionality**, and functionality formulae are presented for discriminators and pyramids - the pyramidal functionality formula presented here being a slight improvement on that independently derived by Al-Alawi & Stohnam (1989a, 1989b).

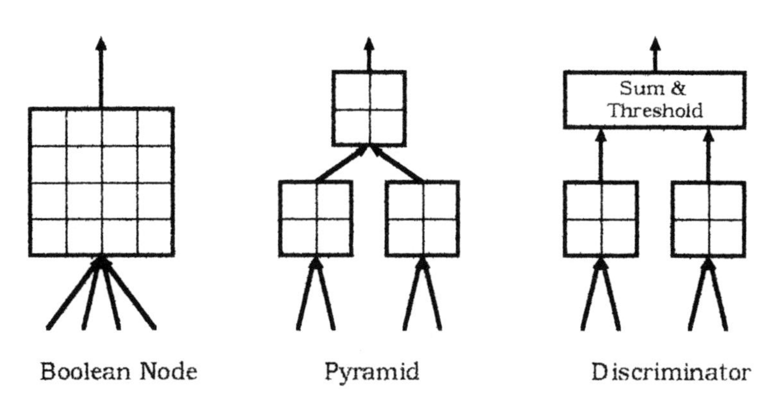

figure 2.2 - decomposing a 4-input boolean node

2.3.2 Generalisation and generalising RAMs

Generalisation is the complement of functionality. It refers the way that training an ANN on one example affects its response to other examples; hence generalisation also depends on the training algorithm used. McCP nodes are not universal and so have generalisation built in. Boolean nodes *are* universal and so generalisation in a network of boolean nodes is a result of the topology of the network - not the nodes themselves. Thus generalisation is seen to be an **emergent property** of the network - the generalisation of the whole is greater than the sum of the generalisations of the parts. At a different level LLNs can be thought of as neuron models with built in generalisation, and the distinction between nodal and topological generalisation in a network of LLNs is again blurred.

A **Generalising RAM, GRAM**, is a PLN with an extra training step, **spreading**, which distributes already learnt information within the PLN (Aleksander, 1990). Spreading is similar to **training with noise**, in which a neural network is trained on an input/ output pair and then trained to give the same output for randomly distorted versions of the input. A typical training and spreading algorithm for a PLN is:

```
For each memory location ω in the PLN                    (initialise)
    Set contents of ω to U
    Unset hiFlag for ω
Train on input / output pairs using, say, Myers' algorithm.
For each memory location ω in the PLN                    (spread)
    If contents of ω is U
        sum := 0
        For each bit in ω
            Make address φ from ω by reversing that bit.
                If not (contents of φ is U or hiFlag for φ set)
                    sum := sum + 2 x contents - 1
        If sum > 0
            Set contents of ω to 1
            Set hiFlag for ω
        If sum < 0
            Set contents of ω to 0
            Set hiFlag for ω
For each memory location ω in the PLN                    (clean up)
    Unset hiFlag for ω
```

The spreading algorithm needs a distance measure, **metric**, on the memory locations - in this case **Hamming distance** is the number of bits different in two addresses, and spreading is to Hamming distance 1. Spreading can be thought of as an explicit training step or else as part of the node function of a GRAM with generalisation built in. In the latter case a PLN can be thought of as a GRAM generalising to Hamming distance 0. In other words a PLN is a special case of a GRAM.

2.3.3 Discriminators

WISARD (Aleksander et al, 1984) is a set of discriminators consisting of RAM's whose outputs are summed. Each RAM is mapped to a group of pixels, **ntuple**, in the retina. During training the patterns seen on the ntuples are stored in the corresponding RAM's. When a test image is put to the retina the discriminators report the number of ntuple patterns that were seen during training. Variations of WISARD can improve discrimination: for example, allowing a count of the number of times an ntuple pattern is seen rather than just logging it seen or not; or weighting the contribution of each RAM to the sum according to its individual discriminative ability, assessed in a second pass of the training set.

The functionality of a thresholded WISARD discriminator is:

$$F = 2 + R.2^{R.2^n} - 2\sum_{m=1}^{R}\sum_{p=m}^{R} C_p^R \left(2^{2^n}-1\right)^{R-p} \tag{2.18}$$

where:

$C_q^p = \frac{p!}{(p-q)!\,q!}$ ways of choosing q items from a set of p items.

R is the number of RAM's in the discriminator.

t is the threshold, $-1 \leq t \leq R$

n is the number of inputs to each RAM, **ntuple size**.

It is assumed that the R n-input nodes are connected to nR independent pixels in the retina - that is the retina is **covered** once.

Discriminators are typically used as classifiers, with one discriminator for each class. Trained on single images they act similarly to Hamming comparators - which are simpler and faster. Trained on multiple images, discriminators and can out-perform Hamming comparators trying to find the average distance from an image to members of a class.

2.3.4 Reversing discriminators

Having trained a discriminator it is possible to **reverse** it to get a class example or set of class examples of what it has learnt. This is particularly useful in unsupervised nets and feature detection/ description from noisy training examples. The aim is to find values of the inputs which cause the discriminator, and therefore its component boolean nodes, to output 1. Partially differentiating (2.3) with respect to the x_i and equating to zero, yields a set of simultaneous equations:

$$\frac{\partial r}{dx_m} = \sum_{\omega \,\varepsilon\, \mathbf{B}^n} a_\omega \frac{v(\omega_m, x_m) - v(1,0)}{x_m\, v(\omega_m, x_m)} \prod_{i=1}^{n} v(\omega_i, x_i) = 0 \tag{2.19}$$

For n=2 this gives solutions:

$$x_1 = \frac{a_{01}}{a_{01} + a_{10} - a_{11} + a_{00}}, \qquad x_2 = \frac{a_{10}}{a_{01} + a_{10} - a_{11} + a_{00}}$$

which may be out of the range [0, 1] e.g. if $a_{00}+a_{11} \geq a_{01}+a_{10}$. In other situations there may be infinitely many solutions or none. Instead we base solutions on:

$$x_m = 2^{-n} \sum_{\omega \in \mathbb{B}^n} a_\omega (2\omega_m - 1) + \frac{1}{2} \tag{2.20}$$

which for n=2 looks like:

$$x_1 = \frac{1}{4}(a_{10} - a_{01} + a_{11} - a_{00}) + \frac{1}{2},$$

$$x_2 = \frac{1}{4}(a_{01} - a_{10} + a_{11} - a_{00}) + \frac{1}{2}$$

Values of x_i close to $\frac{1}{2}$ indicate contradictory features or else features depending on more than n variables. **Constraint contrast enhancement** is a technique whereby the contrast is enhanced by sytematically setting such inputs to 1 or 0 and using them in turn to constrain in other low contrast inputs. It is used in conjunction with a class of complex, high coverage mappings between the discriminator and the retina, and returns a list of high contrast images rather than a single low contrast image.

2.3.5 Pyramids

Boolean nodes are also commonly connected together in tree or **pyramid** structures. In a regular pyramid, with a **fan-in** of n, each bottom-layer node receives its input from the retina and propagates its response to one node in the next layer. Each node in a successive layer is fed by n nodes from the previous layer until one top-level node outputs a binary response for the whole net. Higher band-width responses can be made by having several pyramids, each of which sees the whole input and responds independently.

For a regular pyramid the total number of functions is:

$$F = \sum_{k=0}^{n} C_k^n \, S_{n-k} \, p^{(n-k)} \tag{2.21}$$

where:

$$S_m = \sum_{j=0}^{m} (-1)^j \, C_j^m \, 2^{2(m-j)} \tag{2.22}$$

$p = \begin{cases} (F_s - 2)/2 \text{ if } F_s > 1 \\ 1 \text{ if } F_s = 1 \end{cases}$ the number of non-trivial, F_s the total number functions that sub-pyramid s can perform.

S_{n-k} is the number of functions, given a choice of k inputs, of the top node which are independent of just those k inputs.

The functionality formula is recursive, working down the sub-pyramids until it comes to one of the 'base' cases: a single line, direct to the retina, or a boolean node with n input lines. In the case of a single line there is one non-trivial function: the value on the line itself, so $F_s = 1$ and $p = 1$ (not $(F_s - 2)/2$). For a boolean node with n input lines $F_s = 2^{2^n}$ the number of states that the node can be in. For example in a 2 input node n=2, $F_s = 16$ and p=7.

Because the functionality of each sub-pyramid is calculated, the formula for the functionality as it stands can cope with changes of fan-in between levels. To deal with changes of fan-in within a level - i.e. nodes on the same level have different fan-ins - equation (2.22) is changed so that, instead of multiplying be the combinatorial coefficient C_k^n, a sum is made of all choices of k sub-pyramids and the term $p^{(n-k)}$ is replaced by the product of the p values for the chosen sub-pyramids:

$$F = \sum_{k=0}^{n} S_{n-k} \sum_{\substack{\text{choices } \lambda \\ \text{of } n-k \\ \text{sub-pyramids}}} \prod_{j=1}^{k} p_{\lambda j} \qquad (2.23)$$

where:

$p_{\lambda j} = \begin{cases} (F_{\lambda j} - 2)/2 \text{ if } F_{\lambda j} > 1 \\ 1 \text{ if } F_{\lambda j} = 1 \end{cases}$

$F_{\lambda j}$ is the functionality of the jth sub-pyramid in choice λ.

In the case of a single line, direct to the retina, $F_{\lambda j} = 1$ and $p = 1$ (not $(F_{\lambda j} - 2)/2$).

2.3.6 Networks of LLNs

All McCP networks have their boolean or LLN analogues. Discriminators are comparable to Rosenblatt's perceptron. As mentioned before Martland has demonstrated error back propagation on pyramids of boolean nodes. Analogues of Hopfield nets are Wong and Sherrington's

(1989) sparsely connected networks of PLNs and the fully connected **WIS** auto-associator. The WIS is a set of WISARD discriminators whose outputs are thresholded. There are as many discriminators as retinal pixels so their thresholded outputs form a new binary image which can be fed back as a new retinal image (external feedback). WIS has also demonstrated internal feedback and performed some classification of simple formal grammars. Kohonen networks of **c-discriminators**, containing continuous RAMs, were developed by Ntourntoufis (1990).

Functionality of networks increases explosively with size, and the specific neuron model, and local topology, matter much less than the broad flow of information within a network. For example the mapping from the retina to the inputs of an LLN is usually taken to be random, and feedback a fraction of the input band-width. At large scales it becomes likely that the network will be able to perform the desired function whatever type of node is used, so long as dependent signals are at least connected. For real applications the speed of training and operation, and the ability to create large networks are important. Networks of boolean nodes compare favourably in these respects with networks of McCP nodes.

2.4 Implementing large ANNs using MAGNUS

2.4.1 General neural units

A General Neural Unit, **GNU**, is an internal feedback ANN made of

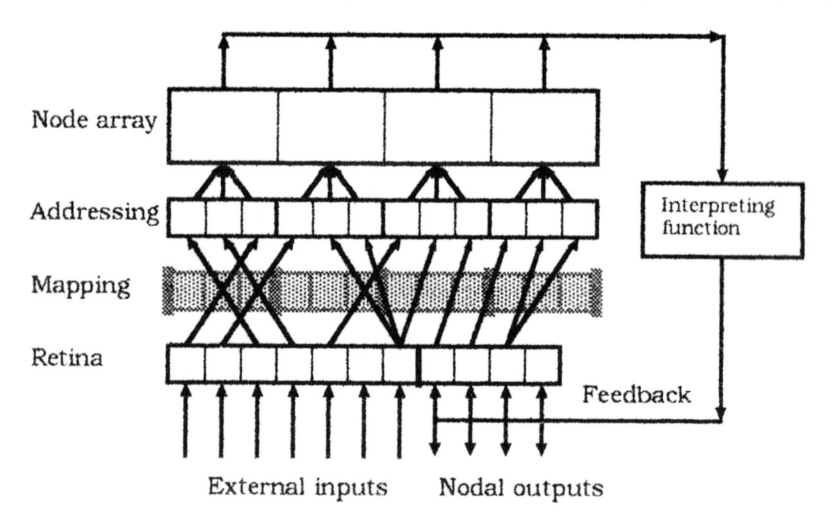

figure 2.3 - representation of a GNU

boolean nodes or LLNs. It can be represented by the diagram (figure 2.3).

This representation is known as **MAGNUS** (Mapping Array GNU Scheme) and it can be used to regularise a variety of boolean ANNs. The topology of the network is a list of offsets to the retina array; it is held in the mapping array. The nodes are taken to be boolean and their outputs are passed through the interpreting function before being fed back into the retina and also output to the user.

Using offsets to a retina containing both externally applied and fed-back signals allows nodes to receive signals of either type or both. Figure 2.4 shows how this scheme could implement a PLN pyramid.

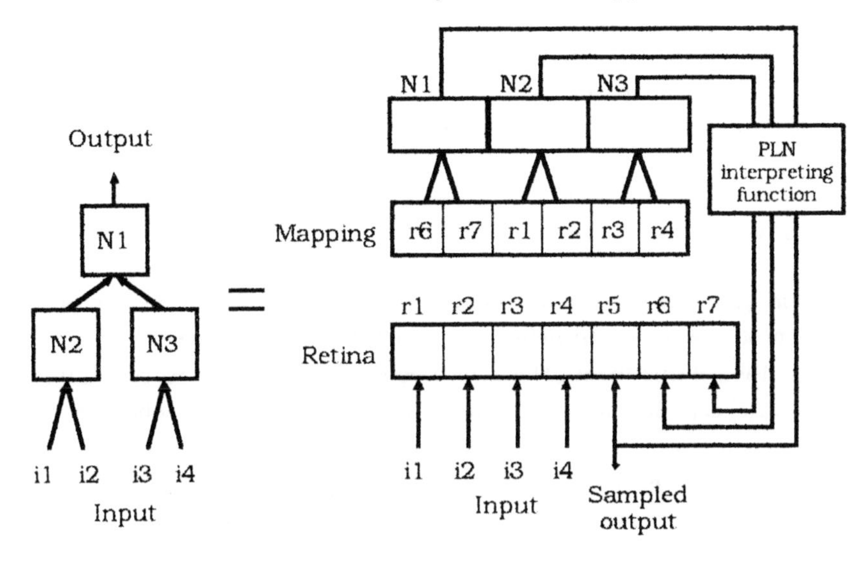

figure 2.4 - implementing a PLN pyramid using MAGNUS

The mapping array is created off-line, although it can be dynamically altered if required. MAGNUS software uses defaults, such as random mappings, and queries the user for any extra information it requires. Thus users have few decisions to make (they are not, however, compelled to use the defaults). With GNUs of GRAMs the only decisions that need be taken are the number of external inputs, the ntuple size of the nodes, and the degree of feedback.

2.4.2 MAGNUS and LLNs

Implementing LLNs or other groupings of nodes requires extra information e.g. number of nodes in the group, ntuple size of nodes within a group, output function (e.g. sum and threshold) of members of a group. Even with these additions the scheme is still very simple and easy to implement efficiently in software, meaning that large networks can be run with minimal overhead and bringing **real applications** (retina sizes ~ 10^4 to 10^5 inputs, 10^3 to 10^4 boolean nodes) within the range of moderately powerful computers.

MAGNUS is being developed as part of the Department of Trade and Industry's IED RUNES project, on which the Neural Systems Engineering Laboratory at Imperial College is working in collaboration with Brunel University and Computer Recognition Systems Ltd of Wokingham. A further aim is to implement a version of MAGNUS on a printed circuit board, **PCB**, to fit in, say, a PC. The software is designed so that few changes should be necessary to take advantage of such a board if it is present.

2.4.3 Conclusion

Function-of-weighted-sum-and-threshold model neurons are so common that definitions of neural networks, and software implementations, may even tacitly assume them. However, boolean model neurons, based upon Random Access Memories, are very different. The simple definition of a neural network as "an interconnected system of parameterised functions" assumes very little, but is useful for building a classification system for neural networks based upon: topology, 'architecture', neuron model, training algorithm and operation schedule.

The original digital neural networks made of RAMs have evolved in two ways: they have become probabilistic, multi-valued and finally, using the address decoder function, analog boolean nodes, and thus equivalent to McCulloch and Pitts nodes with Π-units; and they have been decomposed into Large Logic Neurons - discriminators and pyramids, with limited functionality. Boolean nodes are universal, and in digital neural networks generalisation is a property of the network topology not of the nodes. GRAMs are nodes for which the generalisation may be explicitly defined.

All the major architectures - perceptrons, Hopfield nets, error back-propagation, Kohonen unsupervised learning, even ART1 - have their boolean analogs. Digital neural networks are faster to train and easier to implement in hardware than networks of McCulloch and Pitts nodes, and the MAGNUS scheme for implementing large networks of boolean nodes is being developed so as to reduce processing overhead and facilitate large networks suitable for 'real' applications. It is intended that MAGNUS be implemented on a PCB, and then perhaps on a single chip.

References

Al-Alawi & Stonham (1989a) The functionality of multi-layer boolean neural networks, Electronics Letters, vol 25, No. 10, pp. 657-658

Al-Alawi & Stonham (1989b) A training strategy and functionality analysis of multi-layer boolean neural networks, Brunel University

Wong & Sherrington (1989) Theory of associative memory in randomly connected boolean neural networks. J. Physics A, Math Gen 22, pp 2233-2263

Gorse & Taylor (1988) On the equivalence and properties of noisy neural and probabilistic RAM nets. Physics Letters A, 131 (6)

Gorse & Taylor (1990) Training strategies for probabilistic RAMs. In Parallel Processing in Neural Systems and Computers, Eds. Eckmiller, Hartmann & Hauske, pp. 161-164, Elsevier Science, North Holland

Filho, Bisset & Fairhurst (1990) A goal-seeking neuron for boolean neural networks, Proc. INNC-90-PARIS vol. 2, pp. 894-897, Paris

Kan & Aleksander (1987) A probabilistic logic neuron network for associative learning. Proc. IEEE 1st Ann. Int. Conf. on Neural Networks, San Diego

Myers & Aleksander (1988) Learning algorithms for probabilistic logic nodes. Abstracts of INNS 1st Ann. Meeting p.205, Boston

Myers (1989) Output functions for probabilistic logic nodes. Proc. 1st IEE Int. Conf. on ANNs, pp 163-185, London

Aleksander, Thomas & Bowden (1984) WISARD - a radical step forward in image processing. Sensor Review 4 (3), pp. 29-40. July

Aleksander (1990) Ideal neurons for neural computers. In: Parallel processing in neural systems and computers. Eds. Eckmiller, Hartmann, Hauske. pp. 225-232, North Holland: Elsevier Science

Binstead & Jones (1987) Design techniques for dynamically evolving n-tuple nets. Proc. IEEE vol 134 E, No. 6, November

Rumelhart, Hinton & McClelland (1986) A general framework for parallel distributed processing, Parallel Distributed Processing vol. 1, ch. 2, pp. 45 - 76. MIT

Williams (1986) The logic of activation functions. Parallel Distributed Processing vol. 1, ch. 8, pp. 318 - 362, MIT

Martland (1989) Adaptation of boolean networks using back-error propagation. Abstracts IJCNN-89, p 627, Washington D.C.

Ntourntoufis (1990) Self-organisation properties of a discriminator-based neural network. Proc. IJCNN-90-San Diego, vol 2, pp 319-324

Tank & Hopfield (1987) Collective computation in neuronlike circuits. Sci. Am. December 1987, pp. 62 - 70.

Chapter 3

A survey of neural networks for control

R. Żbikowski and P. J. Gawthrop

3.1 Introduction

This chapter focusses on the feasibility of neural networks and their learning algorithms for identification and control of general nonlinear systems. The most recent learning algorithms for dynamic networks are surveyed with emphasis on the ones relevant to identification of unknown, arbitrary nonlinear plants. The generalisation question is formulated for dynamic systems and discussed from the control viewpoint. A comparative study of stability is made discussing the Cohen-Grossberg and Hopfield approaches. A uniform continuous-time approach to neural networks is presented, as opposed to the dominant discrete-time one. Although the latter results in easy simulations on digital machines, the continuous-time approach gives more insight into the dynamic behaviour of the networks and very fast analog VLSI implementations [14].

3.2 Definitions

Neural networks or connectionist models have been intensively investigated recently [35], [13]. Originally, the inspiration came from the nervous systems of higher organisms, most notably from the human brain [7]. The philosophical, psychological and biological issues which arise are complex and ambiguous and will not be addressed here. From now on a *neural network* or a *connectionist model* will be considered as a computing machine or a dynamical system without any reference to living matter. It is assumed that a neural network is characterised by:

– parallel architecture; it is composed of many self-contained, parallel interconnected processing elements or *neurons*;

– similarity of neurons; each basic processor is described by a standard nonlinear algebraic or differential equation;

– adjustable weights; there are multiplicative parameters each associated with a single interconnection and they are adaptive.

The above structure can be configured in different ways from the signal flow point of view. When the input and intermediate signals are always propagated forwards, the system is called a *static* or *feedforward* network. The flow of information is directed towards the output and no returning paths are allowed (see Fig. 3.1). If either states or outputs are fed back then this is a *dynamic* or *recurrent* network (see Fig. 3.4). The signals are re-used, thus their current values are influenced by the past ones, which is an important characteristic absent in feedforward models.

The key features of neural networks are: nonlinearity, adaptation abilities and parallelism. If we specify a network structure and a learning rule, the network is said to be defined.

As is common in control theory [29] and dynamic systems theory [2], the following general description of finite-dimensional dynamical systems will be assumed:

$$\dot{x} = f(x, u) \tag{3.1}$$

$$y = g(x, u) \tag{3.2}$$

with $x(t_0) = x^0$. Here, $u = u(t), x = x(t), y = y(t)$ are time-dependent vectors and $f(\cdot,\cdot), g(\cdot,\cdot)$ are vector functions of appropriate dimensions. $f(\cdot,\cdot)$ satisfies the Lipschitz condition with respect to x.

3.3 Identification and Control with Static Networks

As mentioned above, there are two general structures of neural networks: static (feedforward) and dynamic (recurrent). This section is devoted to a discussion of the former.

Static networks are widely used for pattern recognition/classification [18] and also for image processing [4]. Another area is approximation theory [28], [8], where the problem may be stated as follows: given two signals $x(t)$ and $y(t)$ find a 'suitable' approximation of the functional relation between the two. This aspect is of interest for us.

3.3.1 Static Networks Description

The most common approach to the approximation problem is to consider a non-dynamic relation

$$y = F(x), \tag{3.3}$$

where y and x are vectors and the parameter t is eliminated. This equation may be approximated with a network having the structure of Fig. 3.1 composed of neurons (as in Fig. 3.2) described by:

$$y_i = \sigma(s_i) + u_i, \quad s_i = \sum_j w_{ij} x_j. \tag{3.4}$$

A common choice of $\sigma(\cdot)$ is a *sigmoid* function

$$\sigma(s) \rightarrow \begin{cases} \rightarrow 1, & \text{if } s \rightarrow +\infty; \\ \rightarrow 0, & \text{if } s \rightarrow -\infty, \end{cases} \tag{3.5}$$

which is usually smooth, but a hardlimiter is also in use [35]. Notice that in Fig. 3.1 all signals are denoted as y_i and numbered from top to bottom. Formula (3.4) and Fig. 3.2 use x_j as inputs to emphasise that the signals fed into static neurons are *not* feedback ones, so not equal to outputs. The reason for introducing the homogeneous notation $y_1, y_2, \ldots, y_N$ will become clear when deriving learning algorithms.

A usual way of arranging the architecture of static networks is to group neurons into *layers* or to design connections in such a way that each of them sends its signal directly only to a certain group of other neurons. As seen in Fig. 3.1, the input layer composed of p distribution nodes (not neurons) feeds the first hidden layer, containing n_1 neurons with outputs $y_{p+1}, \ldots, y_{p+n_1}$. This layer, in turn is connected through weights to the second hidden layer of n_2 units and so on until the last layer of q neurons, called the output layer. Notice that each layer l has a direct connection only to the subsquent one, i.e. $l+1$, making the total of L layers. This results in the description

$$y = \sigma(s) + u, \qquad s = Wy, \tag{3.6}$$

where the matrix W is sparse

$$W = \begin{bmatrix} 0 & 0 & \ldots & 0 & 0 \\ W^1 & 0 & \ldots & 0 & 0 \\ 0 & W^2 & \ldots & 0 & 0 \\ \vdots & \vdots & \ddots & \vdots & 0 \\ 0 & 0 & \ldots & W^L & 0 \end{bmatrix}. \tag{3.7}$$

W^i are matrices of interlayer connection weights. The vectors y and u in (3.6) have the form $y = [y^0 \; y^1 \ldots y^L]^T$ and $u = [u^0 \; u^1 \ldots u^L]^T$, where y^i, u^i are vectors corresponding to the ith layer; u is the constant bias (offset) vector.

3.3.2 Relevance for Control Purposes

Neural networks are adaptive nonlinear systems that adjust their parameters automatically in order to minimise a performance criterion. This obviously links them to adaptive and optimal control. A discussion of these issues will be given in Section 3.3.2. First, the backpropagation algorithm will be presented.

Backpropagation for Static Networks

Backpropagation (BP) was described elsewhere in detail [25], [32], [28], [8]. The reason why it is briefly recalled here is to present it in the rarely used Werbos form [33] and to show later its extensions for dynamic networks. Moreover, the essence of the algorithm gives an immense insight into the differences between static and dynamic networks. Last, but not least, useful notation will be introduced.

Define the ordered system of equations (compare (3.6)–(3.7) and Fig. 3.1)

$$y_i = f_i(y_{i-1}, \ldots, y_1), \qquad i = 1, \ldots, N+1. \tag{3.8}$$

Then a systematic and formally proper way of calculating the partial derivatives of y_{N+1} is as follows:

$$\frac{\partial^+ y_{N+1}}{\partial y_i} = \sum_{j>i}^{N+1} \frac{\partial^+ y_{N+1}}{\partial y_j} \frac{\partial f_j}{\partial y_i}, \tag{3.9}$$

which is a recursive definition of the ordered derivative $\partial^+ y_{N+1}/\partial y_i$, valid only for the systems (3.8).

The network (3.6) is given a reference signal d only for the output layer, i.e. $d_i \neq 0$ for $i = N - q, \ldots, N$ (see Fig. 3.1), and it is required to minimise

$$E = \frac{1}{2} \sum_{t=1}^{P} (d^t - y^t)^T (d^t - y^t), \tag{3.10}$$

which is the nonlinear least-squares fitting problem [5] for P patterns. The gradient algorithm for adjusting weights yields

$$w_{ij}^{new} = w_{ij}^{old} - \alpha \frac{\partial E}{\partial w_{ij}}, \quad \alpha > 0, \tag{3.11}$$

where ($P = 1$ for simplicity; compare (3.10))

$$\frac{\partial E}{\partial w_{ij}} = \frac{\partial E}{\partial y_i} \frac{d y_i}{d s_i} \frac{\partial s_i}{\partial w_{ij}} = \frac{\partial E}{\partial y_i} \sigma'(s_i) y_j. \tag{3.12}$$

For the output layer $\partial E/\partial y_i = y_i - d_i$, but it is not straightforward to find this expression for the hidden layers. The backpropagation hypothesis assumes *linear* propagation of the error derivative

$$\frac{\partial E}{\partial y_i} = \sum_j \frac{\partial E}{\partial y_j} \frac{d y_j}{d s_j} \frac{\partial s_j}{\partial y_i} = \sum_j \frac{\partial E}{\partial y_j} \sigma'(s_j) w_{ji}, \tag{3.13}$$

where y_i belongs to the layer l, and y_j to $l + 1$. Starting from the output layer this can be recursively solved. This formulation requires care with neuron indexing and raises doubts about the conformation with the partial derivative definition.

On the other hand, the network structure is ordered (compare (3.6)–(3.7) and (3.8)). Thus, the hypothesis (3.13) can be expressed as

$$\frac{\partial^+ E}{\partial y_i} = \frac{\partial E}{\partial y_i} + \sum_{j>i}^{N} \frac{\partial^+ E}{\partial y_j} \frac{\partial y_j}{\partial y_i} \tag{3.14}$$

with E treated as y_{N+1} in (3.9). Then the BP algorithm for static networks is

$$\begin{aligned} z_i &= \sum_{j>i} w_{ji} \sigma'(s_j) z_j + \epsilon_i, \\ \frac{\partial E}{\partial w_{ij}} &= \sigma'(s_i) z_i y_j, \\ w_{ij}^{new} &= w_{ij}^{old} - \alpha \frac{\partial E}{\partial w_{ij}}, \end{aligned} \tag{3.15}$$

where $z_i = \partial^+ E/\partial y_i$ and $\epsilon_i = \partial E/\partial y_i$, or in vector-matrix form:

$$S = diag[\sigma'(s_1), \ldots, \sigma'(s_N)],$$

$$z = W^T S z + \epsilon,$$

$$W^{new} = W^{old} - \alpha S z y^T. \tag{3.16}$$

This is inherently an off-line technique as the forward pass must be computed for all t before (3.16) is applied (backward pass).

Both methods lead to the same results for networks of the structure shown in Fig. 3.1 and require only *local* computations. However, Werbos' method corresponds to a fully connected network and thus is more general. Moreover, it is very systematic and straightforward in implementations requiring network architecture modification.

Control Aspects of Static Networks

Neural networks, being nonlinear, do not necessarily offer significant benefits for linear control problems. The real promise of neural networks lies in the realm of nonlinear control. The static approximation capabilities of feedforward networks are of limited use for dynamic system identification and/or adaptive control, since (3.3), unlike (3.1), is an algebraic equation. Difference equations are also algebraic, but recurrent, so it is possible to apply static networks for window of data or off-line only.

As mentioned above, static networks are unable to emulate the differential equation (3.1). This is a problem of substantial importance in the context of adaptive control. Modelling (3.1) for control purposes for a limited workspace through data windowing may be dangerous, because of the rich nonlinear dynamic behaviour of the real plant. In the presence of noise and with possible model mismatch this may lead to instability, especially in the off-line mode (no adaptive updating).

Using static networks for control (as opposed to identification) seems to be more plausible. A static network can be used as a nonlinear gain state feedback controller. However, little is known about how to exploit this possibility, e.g. in the context of feedback linearisation [17], [11], although there are some attempts at this [16]. Learning a system's inverse seems to be a popular approach, but it is limited by definition, sensitive to noise and model uncertainty and suffers from all the drawbacks of static networks emulating dynamic systems, as described above.

3.4 Dynamic Networks for Identification and Control

Dynamic or recurrent networks are qualitatively different from static ones, because their structure incorporates feedback. In general, the output of every neuron is fed back with varying gains (weights) to the inputs of all neurons. A scheme of such a universal neuron is shown in Fig. 3.3. A neural system composed of these elements, with all weights non-zero in general, is called a *fully connected network*. The network architecture is inherently dynamic and usually one-layered, since its complex dynamics gives it powerful representation capabilities. Fig. 3.4 shows a fully connected network configured for control purposes, as explained later. Other characteristics of dynamic networks will be discussed below. Convergence and stability will be analysed after the presentation of learning algorithms.

3.4.1 Mathematical Description

To exhibit dynamic behaviour (3.1) the network must have a dynamic structure. These kinds of networks [9], [10], [21] are built from dynamic neurons (see Fig. 3.3)

$$T_i\dot{y}_i = -y_i + \sigma(s_i) + u_i, \qquad s_i = \sum_{j=1}^{N} w_{ij}y_j \tag{3.17}$$

with $y_i(t_0) = y_i^0$. Here y_j are feedback signals from all neurons and N is the number of neurons in the network.

There are two basic dynamic network descriptions. The first [10] assumes separate state and output, while the other gives linear output equal to state, i.e. in vector matrix form

$$\begin{aligned} s &= Wy, \\ T\dot{y} &= -y + \sigma(s) + u, \end{aligned} \tag{3.18}$$

The matrix W is dense, contrary to that in (3.6)–(3.7), since the network is fully connected. The discrete time version is:

$$\begin{aligned} s(t) &= Wy(t), \\ Ty(t+1) &= \sigma(s(t)) + u(t). \end{aligned} \tag{3.19}$$

We shall use Equations (3.18)–(3.19), which result in less cumbersome formulae.

It should be emphasised that (3.18) and (3.19) describe a *general* model of dynamic networks. The homogeneous notation (all signals denoted as y_i) is used for uniformity and compactness of matrix notation. However, this model should be *configured for control purposes*, which means that some elements of the matrices T and W in (3.18) will be set to 0. The aim is to set up the network in such a way that it approximates the system (3.1). This corresponds to the following design steps (see Fig. 3.4). Input neurons are obtained from (3.17)–(3.18) by setting $T_i = 0$ and $w_{ij} = 0$ for $i = 1, \ldots, p$ and $j = 1, \ldots, N$. For all other neurons ($i = p+1, \ldots, N$) the signals $u_i = 0$. Amongst the neurons different from the input ones, n neurons are picked and their outputs are treated as the *network outputs*. The remaining $\nu > 0$ neurons represent the network's internal *hyperstate*. Thus the network has p inputs u_i, ν hyperstate variables χ_i and n outputs $\hat{x}_i$ making $p+\nu+n = N$ neurons. Note that (3.17)–(3.18) describe *uniform* neurons with their outputs denoted y_i for $i = 1, \ldots, N$. This homogeneity is replaced in Fig. 3.4 by the specialised configuration with neuron outputs denoted according to their function from the control viewpoint.

Recurrent networks possessing the same structure can exhibit different dynamic behaviour, due to the use of distinct learning algorithms. The network is defined when its architecture and learning rule are given. In other words, it is a composition of two dynamic systems: transmission and adjusting systems. The overall input-output behaviour is thus a result of the interaction of both.

There are two general concepts of recurrent structures training. *Fixed point learning* is aimed at making the network reach the prescribed equilibria or perform steady-state matching. The only requirement on the transients is that they die out. *Trajectory learning*, on the other hand, trains the network to follow the desired trajectories in time. In particular, when $t \to \infty$, it will also reach the prescribed steady-state, so it can be viewed as a generalisation of fixed point algorithms.

3.4.2 Fixed Point Learning

For the network (3.18) the error is defined as

$$E = \frac{1}{2} \sum_{k=1}^{P} (y_{\infty}^{k} - y)^{T} (y_{\infty}^{k} - y), \tag{3.20}$$

where y_{∞}^{k} are vectors of the desired equilibria, being solutions of (3.18) with the left-hand side set to 0:

$$0 = -y_{\infty}^{k} + \sigma(s) + u, \quad k = 1, \ldots, P. \tag{3.21}$$

During learning the network does not receive any external inputs. It is excited by initial conditions corresponding to the expected workspace and evolves with y^{∞} as a constant reference signal, which is called *relaxation* [23]. Practically, this is done using *recurrent backpropagation* [22], [1] (with $P = 1$):

$$\begin{aligned} e_i &= \begin{cases} y_i^{\infty} - y_i, & \text{if } y_i \text{ is a network output;} \\ 0, & \text{otherwise,} \end{cases} \\ \dot{z}_i &= -z_i + \sum_j w_{ji} \sigma'(s_j) z_j - e_i, \\ \dot{w}_{ij} &= -\alpha \sigma'(s_i(\infty)) z_i(\infty) y_j(\infty), \end{aligned} \tag{3.22}$$

or, in vector-matrix form

$$\begin{aligned} e &= [0 \,\vdots\, y^{\infty} - y]^{T}, \\ S &= diag[\sigma'(s_1), \ldots, \sigma'(s_N)], \\ \dot{z} &= -z + W^{T} S z - e, \\ \dot{W} &= -\alpha S(\infty) z(\infty) y^{T}(\infty). \end{aligned} \tag{3.23}$$

The calculation of z corresponds to the backpropagation rule for feedforward networks (3.15) with $\epsilon = -e$. The equation for W can be solved after the network has settled, since $y(\infty)$ and $z(\infty)$ are needed. Notice that the right-hand side of the differential equation for W is *constant*, which means that the solution is just the product of the constant and time. Adjustment is thus a one-shot procedure, once the equation for $z(t)$ is solved.

This algorithm differs from Hopfield nets [9], [10], since there the learning was performed by the Hebbian rule [28] or the weights were established by the designer, not the network adaptive algorithm [30]. Thus, there the relaxation is done only during the recall phase (the Hopfield net is a content-addressable memory). Recurrent backpropagation creates basins of attraction around the given equilibria by adjusting the weights, learning them through relaxation (3.23). The Hopfield net generates its attractors and basins by applying fixed weights and thus produces equilibria on its own, representing input patterns or an optimisation problem solution.

If the forward pass in recurrent backpropagation is stable, so is the backward one [1]. On the other hand, the Hopfield net is always stable, because defining it requires guessing a Lyapunov function appropriate to an application [30], [19].

3.4.3 Trajectory Learning

In this case, the error for the network (3.18) is given by

$$E = \frac{1}{2}\int_{t_0}^{t_1}\Big[(d(\tau)-y(\tau))^T(d(\tau)-y(\tau))\Big]d\tau, \tag{3.24}$$

where $d(\tau)$ is a vector of the desired trajectories and t_1 may be a constant (off-line techniques) or a variable (on-line algorithms). The discrete-time version (see (3.19)) yields:

$$E = \frac{1}{2}\sum_{\tau=t_0}^{t_1}\Big[(d(\tau)-y(\tau))^T(d(\tau)-y(\tau))\Big], \tag{3.25}$$

with the previous remarks valid. The case of multiple reference trajectories $d^k(\tau)$ is omitted for simplicity.

Backpropagation Through Time (BPTT)

The simplest method, already mentioned in [25], is to unfold the network through time, i.e. replace a one-layer recurrent network with a feedforward one with t_1 layers. The equivalent static network has as many layers as time instants, which is easily understood in the discrete time context and therefore we start with this. Standard backpropagation is applied, and since each layer contains the same weights "seen" in different moments, their final value is the sum of intermediate values [34]. Thus, we have

$$\begin{aligned}
E &= \frac{1}{2}\sum_{\tau=t_0}^{t_1}\sum_{i=1}^{N}(d_i(\tau)-y_i(\tau))^2, \quad t_1 = const,\\
e_i(\tau) &= \begin{cases} d_i(\tau)-y_i(\tau), & \text{if } y_i(\tau) \text{ is a network output;}\\ 0, & \text{otherwise,}\end{cases}\\
z_i(\tau) &= \sum_{j=1}^{N} w_{ji}(t_0)\sigma'(s_j(\tau+1))z_j(\tau+1) - e_i(\tau), \quad z_i(t_1)=0,\\
\frac{\partial E}{\partial w_{ij}} &= \sum_{\tau=t_0}^{t_1}\sigma'(s_i(\tau))z_i(\tau)y_j(\tau),\\
w_{ij}^{t_1} &= w_{ij}^{t_0} - \alpha\frac{\partial E}{\partial w_{ij}},
\end{aligned} \tag{3.26}$$

where $z_i(\tau) = \partial^+ E/\partial y_i(\tau)$, or, in vector-matrix form,

$$\begin{aligned}
E &= \frac{1}{2}\sum_{\tau=t_0}^{t_1}\Big[(d(\tau)-y(\tau))^T(d(\tau)-y(\tau))\Big], \quad t_1 = const,\\
e(\tau) &= [0 \vdots d(\tau)-y(\tau)]^T,\\
S(\tau) &= diag[\sigma'(s_1(\tau)),\ldots,\sigma'(s_N(\tau))],\\
z(\tau) &= W^T(t_0)S(\tau+1)z(\tau+1)-e(\tau), \quad z(t_1)=0,\\
\nabla_W E &= \sum_{\tau=t_0}^{t_1} S(\tau)z(\tau)y^T(\tau),\\
W^{t_1} &= W^{t_0}-\alpha\nabla_W E,
\end{aligned} \tag{3.27}$$

The definition of $e_i = e_i(\tau)$ in (3.26) and/or $e = e(\tau)$ in (3.27) will be further used without additional explanation. Theoretically, there exists the possibility [38] of an on-line algorithm, but the above calculations would have to be performed every step, requiring unlimited memory and computational power.

Comparison of (3.27) and (3.23) shows that BPTT is a generalisation of recurrent backpropagation. The trajectory to be learned in the fixed-point learning is constant, i.e. $d(\tau) \equiv y^\infty$. The relation between W^{t_1} and W^{t_0} corresponds to the one-shot solution of (3.23).

The continuous time version was introduced heuristically by Pearlmutter [20], and then rediscovered by Sato [27], whose derivation, based on the calculus of variations, is mathematically rigorous and will be briefly presented here. Introduce the Lagrangian

$$L = \int_{t_0}^{t_1}\Big\{\frac{1}{2}\sum_{N-n}^{N}(d_i-y_i)^2 - \sum_{i=1}^{N} z_i[T_i\dot{y}_i + y_i - \sigma(s_i) - u_i]\Big\}d\tau, \tag{3.28}$$

where z_i are Lagrange multipliers (for explanation of index n see Fig. 3.4). The first variation yields

$$\delta L = \int_{t_0}^{t_1}\Big\{\sum_{N-n}^{N}[(y_i-d_i)-z_i]\delta y_i+\sum_{i=1}^{N} z_i\sigma'(s_i)\sum_{j=1}^{N} w_{ij}\delta y_j-\sum_{i=1}^{N} z_iT_i\delta\dot{y}_i+\sum_{i=1}^{N} z_i\sigma'(s_i)\sum_{j=1}^{N} y_j\delta w_{ij}\Big\}d\tau, \tag{3.29}$$

which can be reduced by defining the auxiliary equation

$$T_i\dot{z}_i = z_i - \sum_{j=1}^{N} w_{ji}\sigma'(s_j)z_j + e_i. \tag{3.30}$$

Multiplying both sides of (3.30) by δy_i and summing with respect to i gives

$$\delta L = \sum_{i=1}^{N}[T_iz_i(t_0)\delta y_i(t_0) - T_iz_i(t_1)\delta y_i(t_1)] + \int_{t_0}^{t_1}\Big[\sum_{i=1}^{N} z_i\sigma'(s_i)\sum_{j=1}^{N} y_j\delta w_{ij}\Big]d\tau. \tag{3.31}$$

Since the initial conditions $y_i(t_0)$ do not depend on the weights, $\delta y_i(t_0) = 0$. If additionally the boundary values

$$z_i(t_1) = 0 \tag{3.32}$$

are imposed then

$$\delta L = \int_{t_0}^{t_1} \left[\sum_{i=1}^{N} z_i \sigma'(s_i) \sum_{j=1}^{N} y_j \delta w_{ij} \right] d\tau \tag{3.33}$$

and hence (compare (3.26))

$$\frac{\partial E}{\partial w_{ij}} = \int_{t_0}^{t_1} \sigma'(s_i) z_i y_j d\tau, \tag{3.34}$$

$$\dot{w}_{ij} = -\alpha \frac{\partial E}{\partial w_{ij}}, \tag{3.35}$$

which together with (3.30) completes the algorithm. The vector-matrix form looks as follows:

$$\begin{aligned}
E &= \frac{1}{2} \int_{t_0}^{t_1} \left[(d(\tau) - y(\tau))^T (d(\tau) - y(\tau)) \right] d\tau, \quad t_1 = const, \\
T\dot{z} &= z - W^T S z + e, \\
z(t_1) &= 0, \\
\nabla_W E &= \int_{t_0}^{t_1} S(\tau) z(\tau) y^T(\tau) d\tau, \\
\dot{W} &= -\alpha \nabla_W E.
\end{aligned} \tag{3.36}$$

Notice that the auxiliary equation (3.30) is defined on $[t_0, t_1]$ with $z_i(t_1) = 0$ (see (3.32)), which means that it must be integrated backwards in time. The method for adjusting time constants and delays can be found in [21]. A speeding up modification of the algorithm was proposed in [6].

The method is essentially non-recurrent, because it uses a feedforward equivalent for learning, so there is no explicit use of the network's feedback. It requires recording of the whole trajectories and playing them back to calculate the weights, which are again adjusted according to a one-shot procedure. However, once the weights are established the network runs as a recurrent one and this differentiates it from static backpropagation as described in Section 3.3.2.

Forward Propagation

Backpropagation through time is inherently an off-line technique. The following algorithm [24], [37] overcomes the difficulty. For the recurrent network (T_k suppressed for clarity; compare (3.17))

$$\dot{y}_k = -y_k + \sigma(s_k) + u_k, \quad k = 1, \ldots, N \tag{3.37}$$

with

$$\frac{\partial \dot{y}_k}{\partial w_{ij}} = -\frac{\partial y_k}{\partial w_{ij}} + \sigma'(s_k) \left[\delta_{ik} y_j + \sum_{l=1}^{N} w_{kl} \frac{\partial y_l}{\partial w_{ij}} \right], \tag{3.38}$$

where δ_{ik} is the Kronecker symbol, define error

$$E = \int_{t_0}^{t} \left[\frac{1}{2} \sum_{k=1}^{N} (d_k(\tau) - y_k(\tau))^2 \right] d\tau, \quad t \to \infty, \tag{3.39}$$

i.e. with free termination time (t is a variable). The rationale for introducing the subscript k in (3.37) is that each neuron is fully connected, so it is influenced by all weights w_{ij} (compare (3.38)).

Consider the first variation of the error

$$\delta E = -\sum_{k=1}^{N}\left[\int_{t_0}^{t}[e_k(\tau)\delta y_k]d\tau\right], \tag{3.40}$$

where $e_k(\tau)$ is defined as in (3.26). Since the w_{ij}'s are independent variables, the following holds

$$\delta y_k = \sum_{i=1}^{N}\sum_{j=1}^{N}\frac{\partial y_k}{\partial w_{ij}}\delta w_{ij}. \tag{3.41}$$

Taking into account constraints (3.37) gives, with $p_{ij}^k = \partial y_k/\partial w_{ij}$ substituted to (3.38),

$$\begin{aligned}
\frac{\partial E}{\partial w_{ij}} &= -\sum_{k=1}^{N}\left[\int_{t_0}^{t}[e_k(\tau)p_{ij}^k(\tau)]d\tau\right],\\
\dot{p}_{ij}^k &= -p_{ij}^k + \sigma'(s_k)\left[\delta_{ik}y_j + \sum_{l=1}^{N} w_{kl}p_{ij}^l\right],\\
\dot{w}_{ij} &= -\alpha\frac{\partial E}{\partial w_{ij}},
\end{aligned} \tag{3.42}$$

and

$$p_{ij}^k(t_0) = 0, \tag{3.43}$$

because the initial conditions $y_k(t_0)$ do not depend on weights. The replacement of the boundary values (3.32) with the initial conditions (3.43) allows on-line calculations in (3.42). It is not straightforward to write down (3.42)–(3.43) in the vector-matrix form, because variables p_{ij}^k require a 3-dimensional array. One possibility is:

$$\begin{aligned}
E &= \frac{1}{2}\int_{t_0}^{t}\left[(d(\tau)-y(\tau))^T(d(\tau)-y(\tau))\right]d\tau, \quad t\to\infty,\\
\nabla_W E &= -\sum_{k=1}^{N}\left[\int_{t_0}^{t}[e_k(\tau)P_k(\tau)]d\tau\right],\\
\dot{P}_k &= -P_k + \sigma'(s_k)[\lambda^k y^T + \Pi_k], \quad k=1,\ldots,N,\\
P_k(t_0) &= 0, \quad k=1,\ldots,N,\\
\dot{W} &= -\alpha\nabla_W E,
\end{aligned} \tag{3.44}$$

where $P_k = [p_{ij}^k]$ and $\Pi_k = [\pi_{ij}]$, $\pi_{ij} = \sum_{l=1}^{N} w_{kl}p_{ij}^l$ are square matrices of dimension N and $\lambda^k = [0,\ldots,\underbrace{1}_{\lambda_k},\ldots,0]^T$ is a vector.

A modification of the method can be found in [39].

Forward propagation, called also *real-time recurrent learning (RTRL)* [36], requires non-local computations, since to calculate $\partial E/\partial w_{ij}$ knowledge of all p_{ij}^k, $k =$

$1, \ldots, N$ is needed (see (3.42)). Real-time solving of the auxiliary equation on p_{ij}^k contributes to computing complexity, but the algorithm is truly on-line.

It is worth noting that dynamic networks allow an elegant solution of the key problem of multilayer static networks, viz. the reference signal for hidden units. The answer is straightforward (see (3.26)) by setting the error to 0. It does not make impossible calculation of weights for hidden units (compare (3.42)), since through *feedback* they get information about the system performance.

3.4.4 Control Interpretation of Dynamic Networks

As mentioned in Section 3.3.2 the most promising area of neural networks applications is nonlinear control. If the plant is time-invariant and its description is well-known then feedback linearisation [17], [11] gives a systematic method of controller design. It is, however, sensitive to noise [29] and assumes the canonical model:

$$\dot{x} = f(x) + h(x)u, \tag{3.45}$$

which, although fairly general, is not universal. Sliding control [31] offers some robustness, but requires great skill in guessing an appropriate Lyapunov function. Adaptive nonlinear control [26], [29] is still in its infancy and the describing function method is very limited. Thus, when faced with model uncertainty and real-world noise/disturbances, it is not trivial to design a working controller for a nonlinear plant.

In this context, dynamic neural networks offer an interesting alternative to the conventional approach. They give a generic model for the broadest class of systems considered in control theory. This is an important feature, since the power of *linear* control lies, among other reasons, in the universal method of description. Comparison of the plant state equation and the dynamic network model

$$\begin{aligned} \dot{x} &= f(x,u) && \text{for plant;} \\ T\dot{y} &= -y + \sigma(s) + u && \text{for general dynamic network,} \end{aligned} \tag{3.46}$$

where $x \in \mathcal{R}^n, u \in \mathcal{R}^p$ and $y \in \mathcal{R}^N$, may give the false impression that the network is not more general than (3.45). However, if p neurons (3.17) are chosen with $T_i = 0, w_{ij} = 0$ for $i = 1, \ldots, p,\ j = 1, \ldots, N$ then their outputs are equal to the inputs u_i and are distributed *through weights* w_{ji} to the remaining, unreduced neurons (see Fig. 3.4). Massive parallelism of the network (3.18) is not only introduced to speed it up in a hardware implementation. It also plays the role of plant representation through the network's internal *hyperstate* (compare Fig. 3.4). To avoid the false impression that plant's state $x \in \mathcal{R}^n$ and network's hyperstate $\chi \in \mathcal{R}^\nu$ are equivalent $(n \neq \nu)$, x and χ were introduced, respectively. The homogeneous notation y (see (3.46)) as against the structure specific $u, \chi, \hat{x}$ in Fig. 3.4 was discussed in Section 3.4.1. The network output is deliberately denoted as $\hat{x}$ to emphasise that the network is supposed to output an estimate of the plant's state x. Notice that the network has the same number of inputs, p, as the system and the dimension of its output space (in the network's control configuration in Fig. 3.4) is equal to the dimension of the original state space (i.e. n) or, in other words, $\hat{x} \in \mathcal{R}^n$. It is obvious that the right-hand side of (3.18) cannot be of the same dimension as $f(\cdot,\cdot)$ in (3.1), because it has an entirely different structure. The approach presented here does not necessarily assume that the network is composed of n integrators and a large static network attached to them. This straightforward method reduces the problem to the approximation of $f(\cdot,\cdot)$ within an n-dimensional dynamic architecture. However, it is not obvious whether the considerations about static networks approximation capabilities can be extended to this intuitively appealing architecture. The reason for this is that all finite dimensional plant representations are estimates of systems real dynamics. Therefore, introducing the general model (3.18) with $\nu > 0$ (see Fig. 3.4) a non-minimal realisation of the plant (3.1) is made, which is possibly more robust against unmodelled dynamics or model mismatch.

3.5 Stability of Dynamic Networks

It is characteristic of the enthusiastic approach to neural networks that the central issue of dynamic networks stability did not receive as much attention as simulations and experiments. The most powerful result, the Cohen-Grossberg Theorem [3], was obtained under conservative assumptions (see Section 3.5.1), often violated by practitioners without much harm [22]. Stability of the most popular dynamic architecture (which was used throughout this paper), the Hopfield network [10], is obtained (as was the Cohen-Grossberg Theorem) via Lyapunov's direct method. However, as discussed below, the derivation is not complete.

3.5.1 Cohen-Grossberg Theorem

This result [3] is the most fundamental one in dynamic networks stability theory. It applies to the general model

$$\dot{y}_i = a_i(y_i)[b_i(y_i) - \sum_{j=1}^{N} w_{ij}\sigma_j(y_j)], \quad i = 1,\ldots,N. \tag{3.47}$$

The following assumptions are made for $i, j = 1,\ldots,N$:
1° $w_{ij} \geq 0 \ \wedge \ w_{ij} = w_{ji}$;
2° $a_i(\xi)$ is continuous for $\xi \geq 0$ and positive for $\xi > 0$;
3° $b_i(\xi)$ is continuous;
4° $\sigma_i(\xi) \geq 0$ for $\xi \in (-\infty, +\infty)$, differentiable and monotone non-decreasing for $\xi \geq 0$;
5° $\limsup_{\xi\to\infty}[b_i(\xi) - w_{ii}\sigma_i(\xi)] < 0$;
6° either
a) $\lim_{\xi\to 0^+} b_i(\xi) = \infty$
or
b) $\lim_{\xi\to 0^+} b_i(\xi) < \infty$ and $\int_0^\varepsilon \frac{d\xi}{a_i(\xi)} = \infty$ for some $\varepsilon > 0$.
The assumption of weights symmetry, 1°, is the most common in dynamic networks theory (see Section 3.5.2), but here it is augumented with the requirement of non-negativity of w_{ij}. The continuity assumptions 2°–3° and positivity of $a_i(\cdot)$ for positive argument are needed together to prove that if the system (3.47) starts from *positive* initial conditions then its trajectories remain positive. Therefore all

considerations are limited to the positive orthant $\mathcal{R}_+^N$, which together with the non-negativity constraint on w_{ij} is fairly stringent. However, the assumptions about network nonlinearity, 4°, are mild from the practical point of view, unless the hardlimiter is used.

The Lyapunov function (consider $1^o, 4^o$ and positivity of trajectories) for the system (3.47) is

$$V(y) = -\sum_{i=1}^{N} \int_0^{y_i} b_i(\xi)\sigma_i'(\xi)d\xi + \frac{1}{2}\sum_{i=1}^{N}\sum_{j=1}^{N} w_{ij}\sigma_i(y_i)\sigma_j(y_j) \tag{3.48}$$

and has the time derivative $\dot{V}(y) \leq 0$ of the form

$$\dot{V}(y) = -\sum_{i=1}^{N} a_i(y_i)\sigma_i'(y_i)[b_i(y_i) - \sum_{j=1}^{N} w_{ij}\sigma_j(y_j)]^2. \tag{3.49}$$

Notice that in (3.49) assumptions 2° and 4° were used.

The generality of the theorem is achieved at the cost of a variety of technical (5°–6°) and strong (1°) conditions, which are often violated in practice. The violations do not always lead to instability, because the theorem gives only a sufficient condition.

3.5.2 Hopfield Nets Stability

Hopfield published two fundamental papers [9], [10] on certain network architectures, later called Hopfield nets. However, the structure in [9] differs substantially from that of [10]. We shall concentrate here on the continuous Hopfield net [10].

The name Hopfield net is widely used for the continuous model described by

$$\begin{aligned} T_i\dot{x}_i &= -x_i + \sum_{j=1}^{N} w_{ij}y_j + u_i, \\ y_i &= \sigma_i(x_i), \qquad i = 1, \ldots, N \end{aligned} \tag{3.50}$$

where $\sigma(\cdot)$ is a sigmoid function (3.5). These equations were mentioned when (3.18) was introduced. The Lyapunov function candidate proposed by Hopfield is

$$E = -\frac{1}{2}\sum_{i=1}^{N}\sum_{j=1}^{N} w_{ij}y_iy_j + \sum_{i=1}^{N} \rho_i \int_0^{y_i} \sigma_i^{-1}(\xi)d\xi - \sum_{i=1}^{N} u_iy_i, \tag{3.51}$$

where $\rho_i > 0$ are constants, $\sigma(\cdot)$ is monotone increasing and $w_{ij} = w_{ji} \;\; \forall i, j$. It is straightforward to calculate that $\dot{E} \leq 0$.

According to Hopfield w_{ij} can be of both signs (or 0), which results in the indefiniteness of W in the general case. Neglected discussion of positivity conditions for (3.51) in [10] was addressed in [15] and worked out in detail, showing which functions of the type (3.51) will lead to stable designs. Although through a simple transformation (3.50) can be obtained from (3.47) it does not remove the strong assumptions associated with the Cohen-Grossberg Theorem [12].

As in the discrete model, the continuous Hopfield net does not give a method of weights adjusting (compare Section 3.4.2) — on the contrary it assumes that they are given beforehand and are symmetric, usually with $trW = 0$. Lyapunov functions similar to (3.51) are used to obtain such w_{ij} [30].

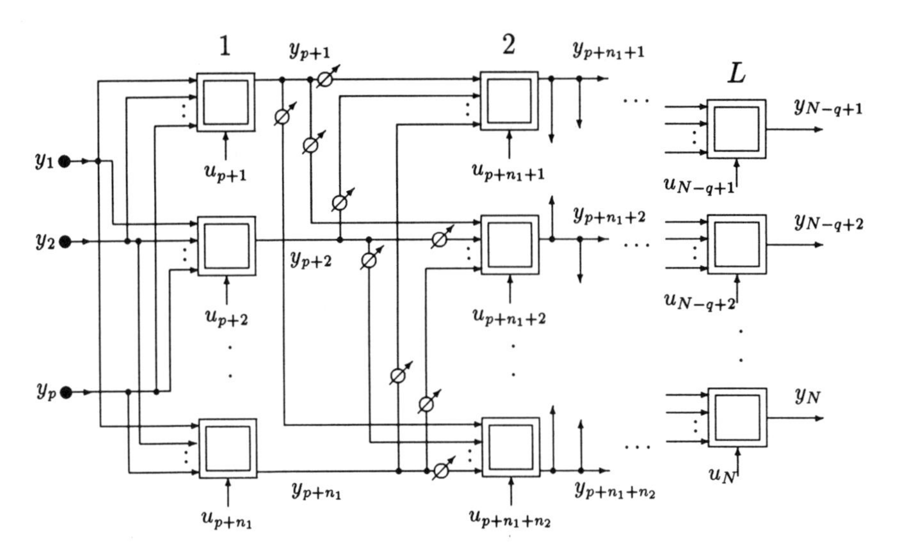

Figure 3.1: Feedforward neural network.

3.6 Conclusions

Neural networks are nonlinear, adaptive and interconnected large-scale systems having two general classes: static and dynamic.

The static one is described by a set of algebraic equations and cannot emulate systems dynamics, which is needed for identification and/or adaptive control. However, static networks possess proven capabilities of arbitrary nonlinear function approximation which can be exploited in the context of state feedback control.

Dynamic networks are universal, parametric, nonlinear dynamic systems of possible use as identifiers or dynamic controllers. The complexity of analysis and synthesis involved is a profound theoretical challenge. However, dynamic continuous networks offer a universal and powerful approach to nonlinear adaptive control as an alternative to the limitations of existing nonlinear control techniques.

Acknowledgements

Rafał Żbikowski's work was supported by the EC TEMPUS Grant no. IMG–PLS–0630–90. Peter J. Gawthrop is Wylie Professor of Control Engineering.

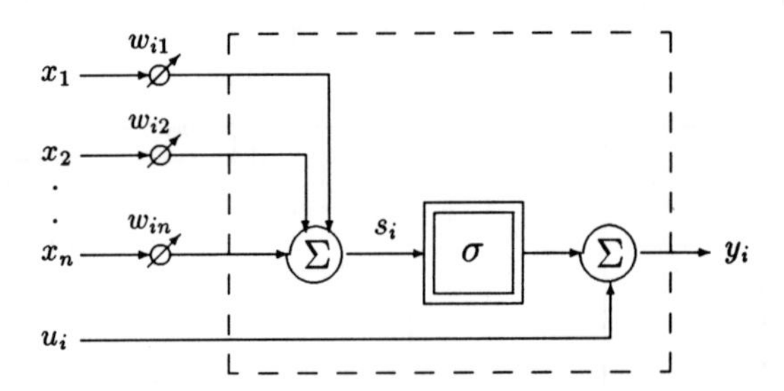

Figure 3.2: Static neuron. The frame stands for a double square in Fig. 3.1.

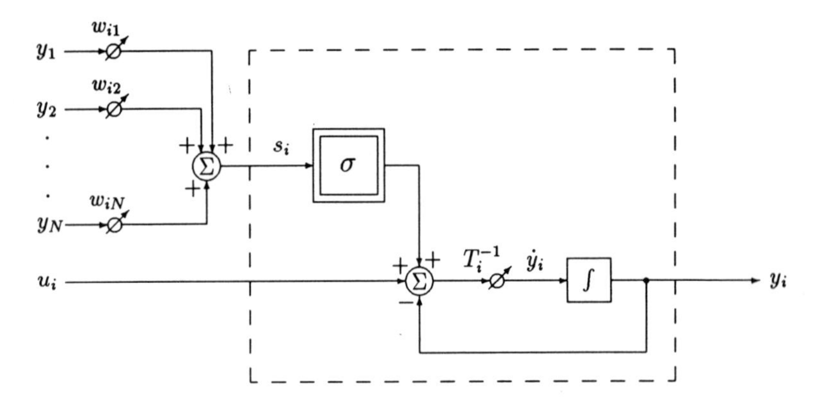

Figure 3.3: Dynamic neuron. The frame stands fcr a double square in Fig. 3.4.

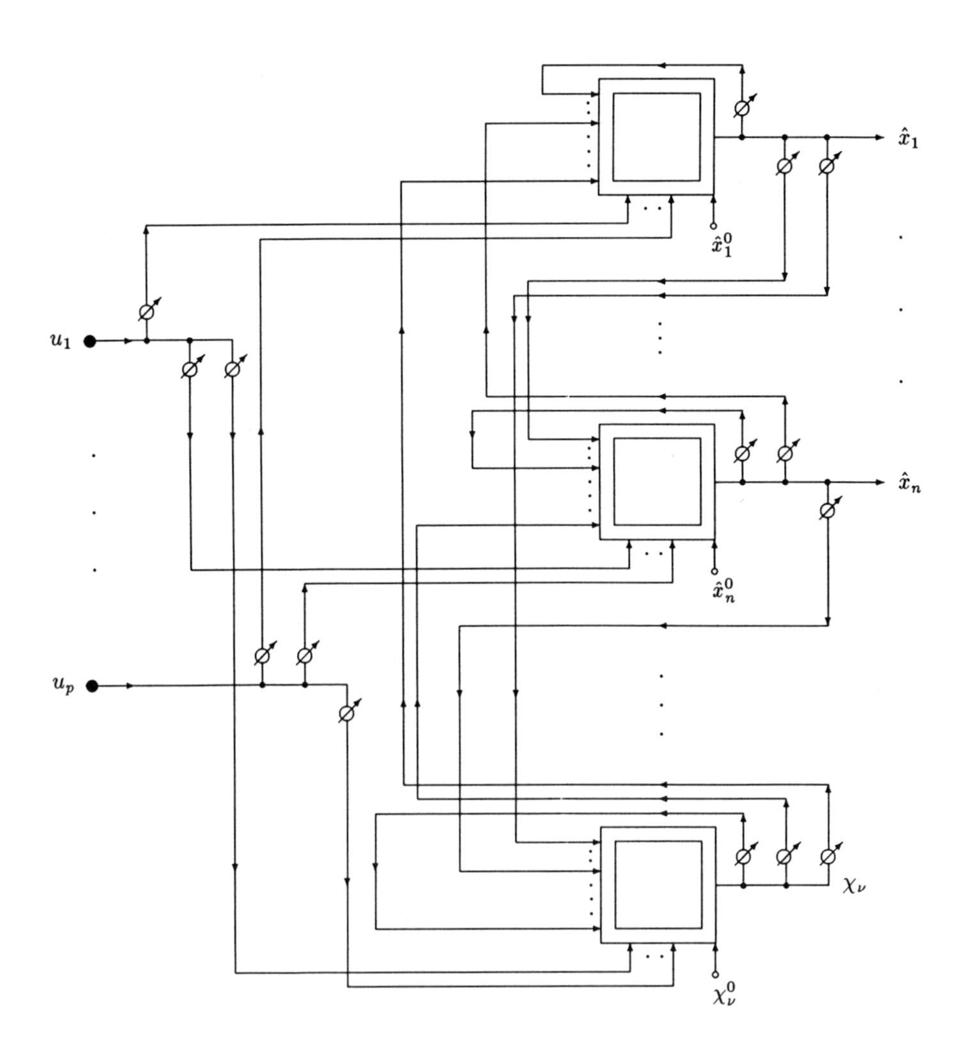

Figure 3.4: Recurrent neural network configured for control purposes.

Bibliography

[1] L.B. Almeida. Backpropagation in Perceptrons with Feedback. In *NATO ASI Series, Vol. F41, Neural Computers, Edited by R. Eckmiller and Ch. v. d. Malsburg*, 1988.

[2] D.V. Anosov and V.I. Arnold. *Dynamical Systems I.* Springer-Verlag, Berlin, 1988.

[3] M.A. Cohen and S. Grossberg. Stability of Global Pattern Formation and Parallel Memory Storage by Competitive Neural Networks. *IEEE Trans. on Systems, Man, and Cybernetics*, 13:815–826, 1983.

[4] A. Dzieliński, S. Skoneczny, R. Żbikowski, and S. Kukliński. Cellular Neural Network Application to Moiré Pattern Filtering. In *Proc. IEEE CNNA'90 Workshop, Budapest, Hungary*, 1990.

[5] P. Eykhoff. *System Identification: Parameter and State Estimation.* Wiley, London, England, 1974.

[6] Y. Fang and T.J. Sejnowski. Faster Learning for Dynamic Recurrent Backpropagation. *Neural Computation*, 2:270–273, 1990.

[7] D.O. Hebb. *The Organization of Behavior.* Wiley, New York, 1949.

[8] R. Hecht-Nielsen. *Neurocomputing.* Addison-Wesley, Reading, Massachusetts, 1989.

[9] J.J. Hopfield. Neural Networks and Physical Systems with Emergent Collective Computational Abilities. *Proc. Natl. Acad. Sci.*, 79:2554–2558, 1982.

[10] J.J. Hopfield. Neurons With Graded Response Have Collective Computational Properties Like Those of Two-state Neurons. *Proc. Natl. Acad. Sci.*, 81:3088–3092, 1984.

[11] A. Isidori. *Nonlinear Control Systems: An Introduction. 2nd Ed.* Springer-Verlag, New York, 1989.

[12] J.H. Li, A.N. Michel, and W. Porod. Qualitative Analysis and Synthesis of a Class of Neural Networks. *IEEE Trans. on Circuits and Systems*, 35:976–986, 1988.

[13] R.P. Lippmann. An Introduction to Computing with Neural Nets. *IEEE ASSP Magazine*, pages 4–22, 1987.

[14] C. Mead. *Analog VLSI and Neural Systems.* Addison-Wesley, Reading, Mass., 1989.

[15] A.N. Michel, J.A. Farrell, and W. Porod. Qualitative Analysis of Neural Networks. *IEEE Trans. on Circuits and Systems*, 36:229–243, 1989.

[16] W.T. Miller, R.S. Sutton, and P.J. Werbos. *Neural Networks for Control.* MIT Press, Cambridge, Massachusetts, 1990.

[17] H. Nijmeijer and A. van der Schaft. *Nonlinear Dynamical Control Systems.* Springer-Verlag, New York, 1990.

[18] Y. Pao. *Adaptive Pattern Recognition and Neural Networks.* Addison-Wesley, Reading, Mass., 1989.

[19] J. Park and S. Lee. Neural Computation for Collision-free Path Planning. In *Proc. of IEEE Int. Joint Conf. on Neural Networks, IJCNN'90, Washington D.C., USA*, 1990.

[20] B.A. Pearlmutter. Learning State Space Trajectories in Recurrent Neural Networks. *Neural Computation*, 1:263–269, 1989.

[21] B.A. Pearlmutter. Dynamic Recurrent Neural Networks. Technical Report CMU-CS-90-196, Carnegie Mellon University, School of Computer Science, 1990.

[22] F.J. Pineda. Generalization of Back-Propagation to Recurrent Neural Networks. *Physical Review Letters*, 59:2229–2232, 1987.

[23] F.J. Pineda. Recurrent Backpropagation and the Dynamical Approach to Adaptive Neural Computation. *Neural Computation*, 1:161–172, 1989.

[24] A.J. Robinson and F. Fallside. Static and Dynamic Error Propagation Networks with Application to Speech Coding. In *Proc. of Neural Information Processing Systems. (ed. D.Z. Anderson), American Institute of Physics*, 1987.

[25] D.E. Rumelhart and J.L. McClelland. *Parallel Distributed Processing: Explorations in the Microstructures of Cognition,* Vol. 1: *Foundations.* MIT Press, Cambridge, Mass., 1986.

[26] S. Sastry and M. Bodson. *Adaptive Control. Stability, Convergence, and Robustness.* Prentice-Hall, Englewood Cliffs, 1989.

[27] M. Sato. A Learning Algorithm to Teach Spatiotemporal Patterns to Recurrent Neural Networks. *Biological Cybernetics*, 62:259–263, 1990.

[28] P.K. Simpson. *Artificial Neural Systems.* Pergamon Press, New York, 1989.

[29] J.E. Slotine and W. Li. *Applied Nonlinear Control.* Prentice-Hall, Englewood Cliffs, 1991.

[30] D.W. Tank and J.J. Hopfield. Simple Neural Optimization Networks: An A/D Converter, Signal Decision Circuit, and a Linear Programming Circuit. *IEEE Trans. on Circuits and Systems*, 33:533–541, 1986.

[31] V.I. Utkin. *Sliding Modes and Their Application in Variable Structure Systems.* Mir Publishers, Moscow, 1978.

[32] P.J. Werbos. *Beyond Regression: New Tools for Prediction and Analysis in the Behavior Sciences.* Ph.D. Thesis, Harvard University, Committee on Applied Mathematics, 1974.

[33] P.J. Werbos. Maximizing Long-Term Gas Industry Profits in Two Minutes in Lotus Using Neural Network Methods. *IEEE Trans. on Systems, Man, and Cybernetics*, 19:315–333, 1989.

[34] P.J. Werbos. Backpropagation Through Time: What It Does and How To Do It? *Proceedings of IEEE*, 78:1550–1560, 1990.

[35] B. Widrow and M.A. Lehr. 30 Years of Adaptive Neural Networks: Perceptron, Madaline, and Backpropagation. *Proceedings of IEEE*, 78:1415–1442, 1990.

[36] R.J. Williams and D. Zipser. Experimental Analysis of the Real-time Recurrent Learning Algorithm. *Connection Science*, 1:87–111, 1989.

[37] R.J. Williams and D. Zipser. A Learning Algorithm for Continually Running Fully Recurrent Neural Networks. *Neural Computation*, 1:270–280, 1989.

[38] R.J. Williams and D. Zipser. Gradient-Based Learning Algorithms for Recurrent Connectionist Networks. Technical Report NU-CCS-90-9, Northeastern University, Boston, College of Computer Science, 1990.

[39] D. Zipser. A Subgrouping Strategy that Reduces Complexity and Speeds Up Learning in Recurrent Networks. *Neural Computation*, 1:552–558, 1989.

Chapter 4

Control applications for feed forward networks

G. Lightbody, Q. H. Wu and G. W. Irwin

4.1 Introduction

This chapter illustrates the use of feedforward neural networks, such as the Multilayer Perceptron, for nonlinear adaptive control. Control structures, based on the ability of the feed-forward network to form an arbitrary nonlinear control law, will be proposed along with possible methods by which these nonlinear neural controllers can be trained. Utilising the neural network in this way, it is possible to realise a number of controllers such as a nonlinear extension to the Model Reference Adaptive Controller, (MRAC), nonlinear adaptive regulators or simply fixed-gain nonlinear controllers.

A possible neural MRAC scheme is proposed, applied to the control of a number of simulated plants and compared with conventional, Lyapunov based MRAC. It is shown that, given the restriction of non time-varying plants, once trained a suitable nonlinear controller can be produced. Finally, using a novel approach to solve the problem of adapting the controller parameters, a neural network based adaptive regulation scheme is described. This is applied to the adaptive regulation of a realistic, high order, nonlinear simulation of a turbogenerator, under various faults conditions and found to produce acceptable results.

4.2 Adaptive control

The field of adaptive control developed in response to the need to control systems which were time-varying through gradual degradation of the plant or due to disturbances on the system, [1]. It was thought that, by continuously adapting the control gains of the typical linear controller, more accurate control of nonlinear systems could be provided as the setpoint point was varied over the full nonlinear range. The schematic of a general adaptive control scheme could be as shown in figure 1. There are two loops present in this structure, the control loop and the parameter adjustment loop. The latter attempts to adjust the control gains to force the plant to follow some desired response.

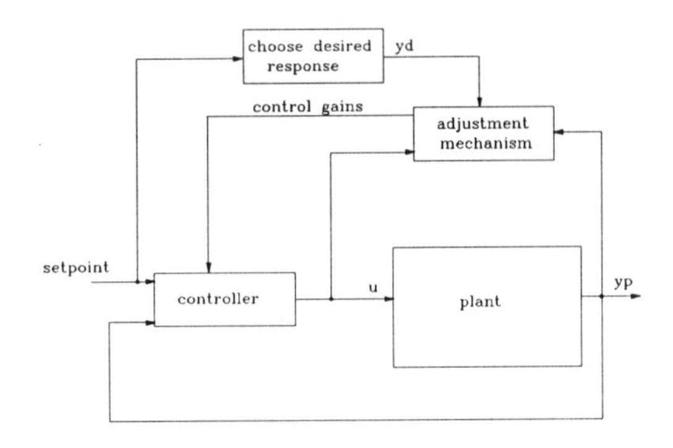

Figure 1 General adaptive control structure

Adaptive controllers can be roughly categorised according to the means by which the controller parameters are adjusted. A common strategy is MRAC, where the controller parameters are changed continuously using the gradient, stability or passivity approaches. Alternatively, self-tuning controllers normally first identify some model of the plant, for example by recursive least squares, then use the parameters of the model to design a suitable controller, by pole-placement or minimum variance for example. All of these methods are usually based on the assumption that the plant can be linearised about a given operating point to produce an approximate linear model. Then, with a linear control law, such a plant can be controlled as desired. However, in practice, industrial systems are highly nonlinear and produce problems with unmodelled dynamics, robustness and the obvious inaccuracy of a control law which generates control as a linear combination of the system states. Hence the need for nonlinear modelling techniques and for nonlinear control.

4.3 Neural networks for adaptive control

Feedforward networks, such as the Multilayer Perceptron consist of simple processing elements arranged in layers. Each element takes as input the weighted sum of all the outputs of the previous layer and passes this through a nonlinear activation function. The structure is feedforward in nature, with no connections within layers or from outer layers towards the input. Training the network involves adjusting the weights, using some learning rule, like backpropagation, so that the network emulates the desired nonlinear mapping from the input to the output vector.

The nonlinear mapping properties of neural networks are central to their use in control engineering. Feedforward networks, such as the Multilayer Perceptron, can be readily thought of as performing an adaptive, nonlinear vector mapping. This can be best understood by recognising that the network carries out the following nonlinear function, where W represents the weights of the network, Y the output vector and X the input vector.

$$Y=F(X,W) \tag{1}$$

Recent theoretical results by Cybenko, [2] and Funahashi, [3], have rigorously proven that a wide range of nonlinear functions can be approximated arbitrarily closely by a feedforward network with only a single hidden layer of nonlinear elements. These are existence proofs and there are no solid theoretical grounds for fixing the number of neurons in the hidden layer or selecting the weights to achieve this mapping.

The feedforward network can therefore be utilised as a general structure for an adaptive nonlinear controller. A possible control scheme based upon such a neural control is shown in figure 2.

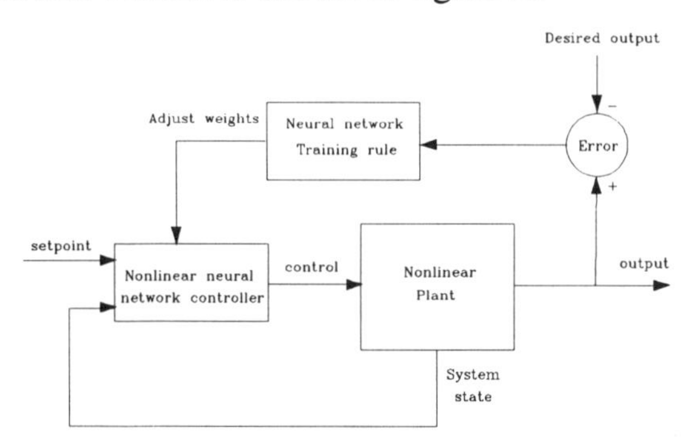

Figure 2 Neural network control scheme

4.3.1 Training techniques for neural adaptive control

For this formulation of the adaptive control problem, it is necessary to adjust the weights of the neural network, during the learning phase, to produce a nonlinear controller, that can control the nonlinear plant in such a manner, as to minimise some cost function of the plant output and the desired response. In linear adaptive control theory, the controller gains are best adjusted using stability techniques, based on Lyapunov or Popov stability theorems, [4]. Such an approach ensures that the gains are adjusted in such a way, so that the error between the plant output and the desired response remains stable and asymptotically reduces to zero. Due to the nonlinearities present in both the plant and the controller, stability based adaptation rules have not as yet been proposed for the neural adaptive controller. The original research into adaptive control was based on gradient descent techniques, such as the MIT rule, in which the cost could be minimised by the adaptation of controller gains in the negative direction of the gradient of the cost with respect to these gains, [1]. A possible adjustment algorithm for the weights of the neural controller could therefore be based on gradient descent, using the backpropagation rule [5] to provide the necessary gradient of the cost function with respect to each weight.

In figure 2, the plant is situated between the neural network and the error, hence it is necessary to find some method by which the error at the output of the plant can be fed back, to produce a suitable descent direction at the output of the neural network. Psaltis and Sideris, [6], introduced the concept of using the plant Jacobian, or sensitivity derivative, to allow errors at the plant output to be fed back to the network. This is not a new concept, the field of sensitivity modelling having been developed to provide estimates of these derivatives, for the use in gradient descent MRAC schemes. If the cost function is defined as J(W), then, knowing the Jacobian of the plant, the gradient of the cost function with respect to the jth input, u_j, could readily be determined, with y_i being the ith plant output.

$$\frac{\partial J(W)}{\partial u_j} = \sum_{i=1}^{n} \frac{\partial J(W)}{\partial y_i} \cdot \frac{\partial y_i}{\partial u_j} \tag{2}$$

This gradient, which is produced at the output of the neural network, can now readily be fed back using the backpropagation training rule. However, since it is to be assumed that little knowledge of the nonlinear plant is available, it would be difficult to obtain an analytical expression for the plant Jacobian. Numerical differentiation could be used to form an approximation to the Jacobian, but would of course suffer from the large errors that plague such a technique. Summarised below, are a number of possible methods by which the problem of backpropagating the errors through the plant to the controller could be solved.

4.3.1.1 Approximation to the plant Jacobian

It was suggested by Saerens and Soquet, [7], to use the sign of the Jacobian, instead of its real value, for the training of neural adaptive controllers. This is often available simply from qualitative knowledge of the system in question. The plant backpropagation equation then becomes:

$$\frac{\partial J(W)}{\partial u_j} \simeq \sum_{i=1}^{n} \frac{\partial J(W)}{\partial y_i}.\mathrm{SGN}\left(\frac{\partial y_i}{\partial u_j}\right) \tag{3}$$

It should be noted that the scalar product between the gradient produced by the true method and that produced by the approximation is always positive and will hence ensure for error minimisation.

4.3.1.2 Back-propagation through a neural plant model

A ready application for feedforward neural networks is in the field of nonlinear system modelling, [8, 9]. The nonlinear vector mapping in equation 1 could easily be reformulated to give the general nonlinear model for a SISO system, with $y_p(k)$ and $u(k)$ being the plant output and input respectively and $W(k)$ the weight vector.

Widrow, [10], and Jordan and Jacobs, [11], proposed using a neural forward model of the plant as a channel for the backpropagation of errors to the neural controller. A neural network is first trained to provide a model of the nonlinear plant in question. This can then be used in parallel with the plant, with errors at the plant output backpropagated through the model to form the necessary gradients at the output of the neural controller, hence avoiding the need to know the plant Jacobian.

Given a neural forward model of the plant is available, Hunt and Sbarbaro, [12], pointed out another interesting variation of this method. Consider that the plant is SISO and approximated by a (N_i, N_h, 1) feedforward neural network to produce the model:

$$\hat{y}_p(k+1) = \hat{f}(\ y_p(k),\ ..,\ y_p(k-p),\ u(k),\ ..,\ u(k-q)\ ) \tag{4}$$

The output of the neural network could be represented by the following series:

$$\begin{aligned} \hat{y}_p(k+1) &= \sum_{i=1}^{N_h} w_{2,i,1} \cdot f_i \\ f_i &= \tanh\left(\frac{1}{2}\left(\sum_{j=1}^{N_i} w_{1,j,i} \cdot x_j + \theta_i\right)\right) \end{aligned} \tag{5}$$

Here, $w_{i,j,k}$ is the weight of the connection between the jth neuron in the ith layer and the kth neuron in the $(i+1)^{th}$ layer, while x_m is the network input at the mth input node. If u(k) is at the pth neural input of this forward model then, from equation 5, the plant Jacobian can be approximated as follows:

$$\frac{\partial y_p(k+1)}{\partial u(k)} \simeq \frac{\partial \hat{y}_p(k+1)}{\partial u(k)} = \frac{1}{2}\sum_{i=1}^{N_h} (1+f_i)(1-f_i) w_{2,i,1} w_{1,p,i} \tag{6}$$

4.3.1.3 Search based optimisation techniques

Bleurer et al., [13], proposed a parameter optimisation approach in which a search technique was used to directly adjust the weights and hence minimise some cost function. In an axis-parallel search one weight of the network is changed at a time to minimise the cost with respect to that weight. Alternatively, for a stochastic search, a random perturbation vector is added to the present weight vector and, if this results in a lower cost, the new weight set would be kept. Otherwise, the present weights are maintained and a new random weight perturbation vector produced. These steps are repeated until the cost function is suitably minimised. Such methods of course are not based on gradient descent or the backpropagation training rule, hence they do not require any knowledge of the plant Jacobian.

4.3.2 A comparison of neural and Lyapunov MRAC schemes

A possible structure for a continuous time, neural MRAC scheme is shown in figure 3.

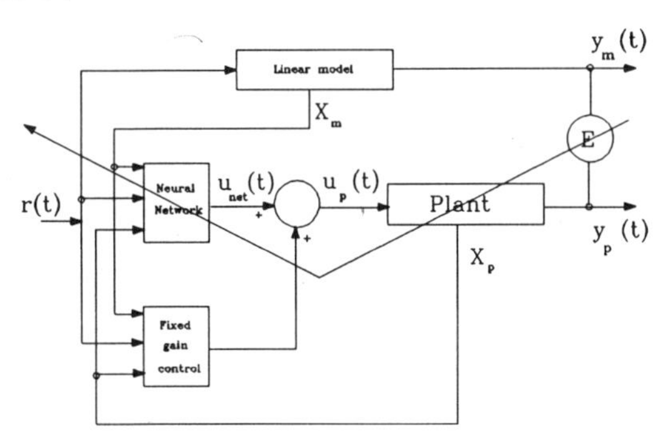

Figure 3 Neural MRAC system

The controller consists of a fixed gain control law in parallel with a neural controller. The fixed gain control is chosen initially to stabilise the plant and to provide approximate control. The neural network then provides a nonlinear element for the control, producing the vector mapping defined in equation 7, where X_p and X_m are the plant and model states respectively and r(t) is the reference signal.

$$u_p = f(X_p, X_m, r(t)) \tag{7}$$

A suitable reference model is chosen to run in parallel with the controlled plant. The output of the model is compared with the plant output to produce an error which, by the backpropagation algorithm, can be fed back to adjust the weights of the neural network and eventually produce convergence of the plant response to that of the desired model. It should be appreciated that the parallel combination of the fixed gain linear controller and the neural network removes the need for a generalised training scheme as discussed by Psaltis et al., [6].

In order to appraise the abilities of neural networks for the adaptive control of nonlinear systems, two simple second-order plants, described by equations 8a and 8b respectively, were chosen as test examples. The first plant is linear for small values of y_p and gradually becomes more nonlinear as y_p approaches 1.0 radian. The highly nonlinear plant (b) is nonlinear over all values of y_p.

$$\begin{aligned} \text{a)} \quad \ddot{y}_p(t) &= 2.0 \sin(y_p(t)) + u_p(t) \\ \text{b)} \quad \ddot{y}_p(t) &= 2.0\, y_p(t) \sin(y_p(t)) + u_p(t) \end{aligned} \tag{8}$$

For simplicity, the sign of the plant Jacobian was used to allow the errors to be backpropagated through the plant. For comparison purposes, results were also obtained from conventional model reference controllers developed using Lyapunov stability theory. Given a setpoint signal r(t) and the output of the neural network controller $u_{net}(t)$, the combined fixed gain and neural control equation for both plants was chosen as:

$$u_p(t) = -3.6\, y_p(t) - 1.6\, \dot{y}_p(t) + r(t) + u_{net}(t) \tag{9}$$

The linear reference model chosen for these plants to follow was:

$$\ddot{y}_m(t) = -\, y_m(t) - 1.6\, \dot{y}_m(t) + r(t) \tag{10}$$

A (5, 11, 1) feedforward network was trained using error backpropagation to produce the nonlinear control law. The cost function to be minimised was simply the square of the instantaneous error:

$$J = (\, y_p(t) - y_m(t)\,)^2 \tag{11}$$

The plots of figures 4a and 5a show the performance of the neural controller for the systems described in equation 8. The plots of figures 4b and 5b, show the response of the traditional Lyapunov designed model reference controllers for these systems. From these results, it is apparent that the conventional method was superior for the slightly nonlinear plant (a). However, for the highly nonlinear plant, (b), the conventional MRAC scheme failed to provide adequate model following. In all cases the neural based adaptive controller was capable of converging to produce excellent model following within forty seconds. One very important point to note is that, once trained, the neural controller should be able to provide a nonlinear control law, that is, the controller parameters can be fixed and the desired model following response still obtained. For the traditional methods, any ability to control the nonlinear plant is derived from continuous adaptation of the controller gains.

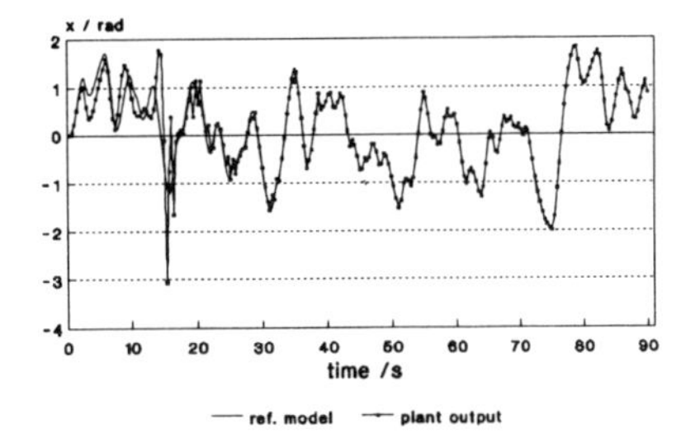

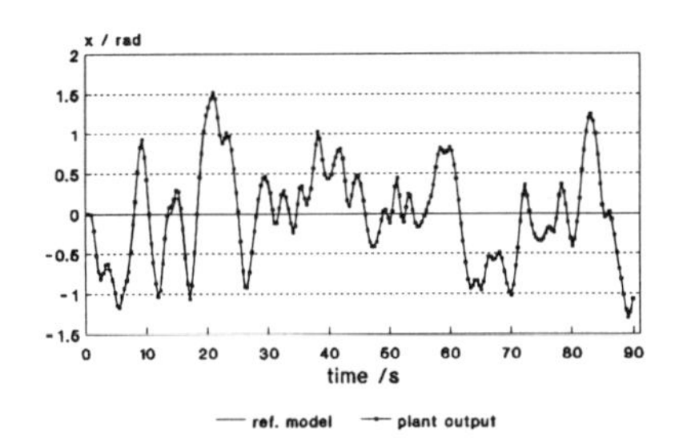

Figures 4a Neural MRAC, plant (a) 4b Lyapunov MRAC, plant (a)

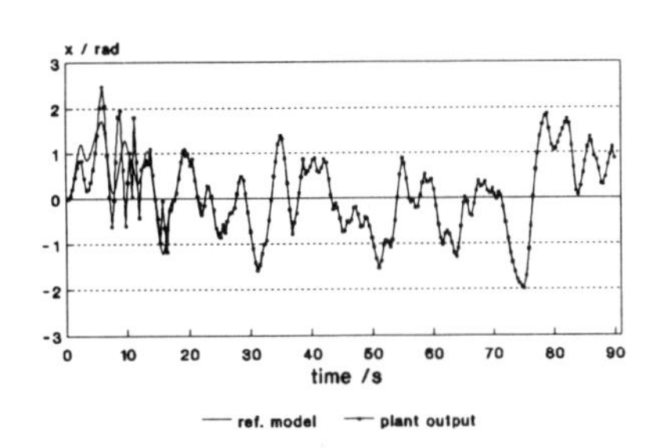

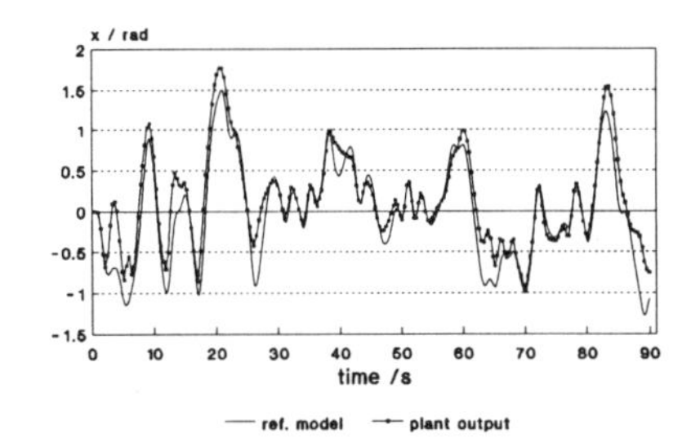

Figures 5a Neural MRAC, plant (b) 5b Lyapunov MRAC, plant (b)

4.4 Turbogenerator neural control - a case study

Turbogenerators are major components in electric power systems and their performance is directly related to security and stability of power system operation. A turbogenerator is a highly nonlinear, fast-acting, multivariable system, connected through a transmission system to a power system. Turbogenerators operate over a wide range of varying operating conditions. Their dynamic characteristics vary as conditions change, but the outputs have to be coordinated so as to satisfy the requirements of power system operation.

Various techniques have been developed to design controllers for unknown turbogenerator systems [14, 15]. Most adaptive control algorithms are designed using linear models, with some strong assumptions of types of noise and disturbances. Based on those models, traditional techniques of identification, system analysis and synthesis can be applied to achieve the desired performance. However, the turbogenerator system is nonlinear, with complex dynamic and transient processes, hence it cannot be adequately described by such linear models. Likewise, for the design of adaptive controllers, it has to be assumed that the number of system inputs equals the number of system outputs. If necessary this could be achieved by using a transformation to reduce the dimensions of the output space, with the drawback that this will degrade the description of the system dynamics. Consequently, the issues of unmodelled dynamics and robustness arise in practical applications of these adaptive control algorithms and hence supervisory control is required, [15, 1].

Recently there has been increasing interest in applying neural networks to power systems but little work has been reported on using such networks for real-time control. This section presents results on the application of neural networks to turbogenerator system control.

A number of specific difficulties arise in the design of a neural network based regulator for the turbogenerator. First a reference model, (as proposed by Narendra, [16]), must be avoided due to difficulties in choosing the reference model for such a complex system. Likewise, the use of an inverse model, as employed by Bassi and Bekely, [17], would be impossible due to high gain loops between the turbogenerator and actuators. Since the turbogenerator is a multivariable system, the conventional neural network based model structure, proposed by Chen, [18], is also unsuitable. To overcome these difficulties and to provide suitable turbogenerator control, a hierarchical architecture was adopted for the neural network regulator. This is similar to the approach used by Nguyen and Widrow, [10], for their truck-backer problem. The neural network regulator consists of two sub-networks which would be used for dynamic modelling and control respectively. This network structure would solve the problem of backpropagating errors at the plant output, across the plant to the neural controller, during the training procedure.

In this case study, the dynamic modelling of a detailed simulation of a turbogenerator system was first investigated. The operation and behaviour of the neural network regulator, based on this dynamic model was then evaluated under different operation conditions and disturbances.

4.4.1 Dynamic modelling

The turbogenerator unit consisted of a synchronous generator and exciter with an automatic voltage regulator and a turbine and governor system. The synchronous generator was represented by a seventh-order, nonlinear differential equation. The excitation system was described as a first-order system, representing a static exciter with a small time-constant and with positive and negative limits on the field voltage. A three-stage turbine, with reheat, was employed to drive the synchronous generator. A constant steam source was assumed. An outline of a turbogenerator, connected to an infinite bus power system, is illustrated in Figure 6. The simulation equations and parameters of the synchronous generator, exciter, turbine and governor systems are contained in reference [14].

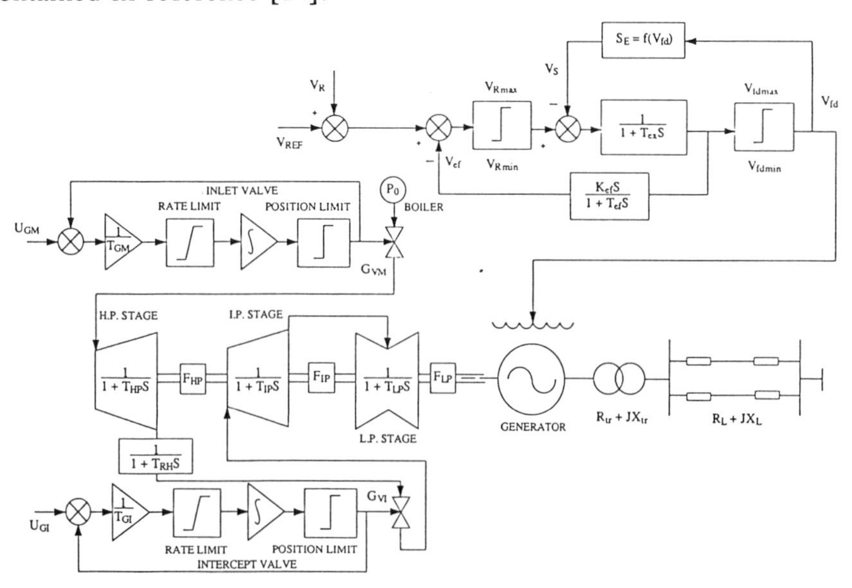

Figure 6 The turbogenerator system

The turbogenerator can be represented by the discrete-time, nonlinear, neural network model shown below:

$$\hat{Y}(k+1) = \hat{F}(\ Y(k),\ \ldots,\ Y(k-p),\ U(k),\ \ldots,\ U(k-q)\ ) \qquad (12)$$

Here $Y(k) \in R^n$ is the output vector, $U(k) \in R^m$ is the input vector and p, q are the orders of the time series $\{Y(k)\}$ and $\{U(k)\}$. $\hat{Y} \in R^n$ is the neural prediction of the output vector. The turbogenerator can then be dynamically modelled by adjusting the network weight vector W(k), so as to minimise the following model cost function.

$$J_m(k) = \frac{1}{2T}\left(\sum_{i=k-T+1}^{k} \sum_{j=1}^{n} r_j \, [y_j(i) - \hat{y}_j(i)]^2 \right) \tag{13}$$

where y_j, $\hat{y}_j$ denote the jth element of the system output Y(t) and the neural network output $\hat{Y}(t)$ respectively. The penalty coefficient r_j is applied on the jth output, so as to coordinate the performance of the system outputs. The time T is the period over which the mapping performance is to be evaluated. Backpropogation was used to provide the necessary gradient vector of the cost function with respect to the weights and the network weight vector is then updated as follows:

$$W(k) = W(k-1) - \frac{\lambda}{T} \sum_{i=k-T+1}^{k} \nabla J_m(W, i) \tag{14}$$

Here $\nabla J_m(W,k)$ is the gradient vector at time k, consisting of the derivatives of J_m with respect to each weight in the network and λ is the step length. Inherent in equation 14 is the averaging of the gradient vector over the period time T. Also, given the fact that the cost of equation 13 is an average cost over a period of time, the training process for the neural network should be more robust, with the speed of convergence of the weights increased.

To control the major outputs of the turbogenerator, the terminal voltage V_t and generator speed ω are measured on-line and used as feedback variables to yield the excitation control signal V_R and governor input U_G. As part of the neural regulation scheme, a dynamic model of the turbogenerator system was produced, of the form defined in equation 12, with the system inputs and outputs chosen as:

$$Y(k) = [\Delta V_t(k), \Delta\omega(k)]^T \qquad U(k) = [\Delta V_R(k), \Delta U_G(k)]^T \tag{15}$$

Here $\Delta V_t(k)$ is the derivation from the setpoint and $\Delta\omega(k)$ is the derivation from the synchronous speed. $\Delta V_R(k)$ and $\Delta U_G(k)$ are the changes in excitation and governor inputs respectively.

A (12, 14, 2) network was employed with the nonlinear activation function y = $(1-e^{-x})/(1+e^{-x})$ for the hidden units and linear input and output elements. The turbogenerator system outputs were sampled every 20 ms. The inputs to the network consisted of samples of Y(t) and U(t) and the orders p, q in equation 12 were both 2. The network weights were initially set as random values in the range [-0.1, 0.1] and updated according to equation 14 with a step length λ of 0.2 and the penalty coefficients applied to the voltage and speed set at 20 and 1 respectively. The averaging period T was chosen to be 10 samples. The turbogenerator system was initialized with the real and reactive powers being, P_t = 0.8 p.u. and Q_t = 0.2 p.u. Two PRBS signals were then employed to excite the simulated turbogenerator system, with amplitudes set at 30% and 20% of the given inputs of the excitation and governor systems respectively. The response shown in figure 7 shows the performance of the neural dynamic modeller, after 200 s training.

Figure 7 shows that neural networks can be used for the dynamic modelling of turbogenerator systems. It should be noted that good mapping quality was achieved even under large disturbances which forced the system to operate well outside the linear range. Thus the neural network based dynamic modeller can in fact realize complex, nonlinear, dynamics whose rich modes are excited by PRBS signals.

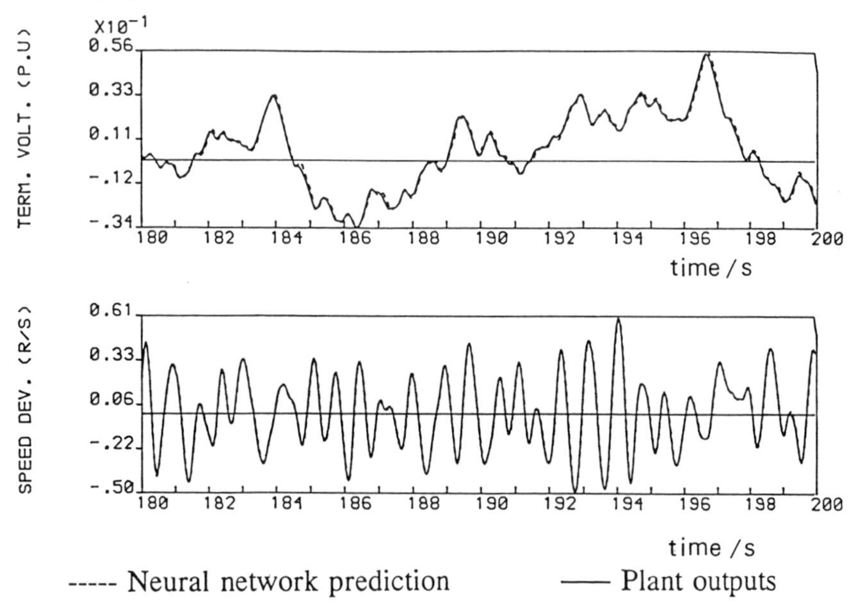

----- Neural network prediction —— Plant outputs

Figure 7 Terminal voltage and speed deviation predictions

4.4.2 Regulator design

The feedback control for the turbogenerator system may be defined as:

$$U(k) = G[Y(k), Y(k-1), \ldots, Y(k-s), U(k-1), \ldots, U(k-r)] \quad (16)$$

The task is then to chose some nonlinear vector function G(.) which will minimise the following cost function:

$$J_c = \frac{1}{2T_c} \sum_{i=k-T_c+1}^{k} \sum_{j=1}^{n} r_j \, (y_j(i) - d_j(i))^2 \quad (17)$$

Where $d_j(i)$ denotes the j-th element of the desired system output D(i) and s,r are the orders of the time series of the feedback variable Y(k) and U(k) respectively. A Multilayer Perceptron can now be trained to provide the nonlinear controller, G(.). However, the cost function of equation 17, depends on the errors between the desired track and the plant output at the actual output of the plant. In order to obtain the gradients of this cost function with respect to the controller weight vector, some method of backpropagating the errors through the system to the plant input is required.

To overcome this problem a hybrid neural network regulator was developed [19]. This consists of two sub-networks, a neural network mapper (NNM) and a neural network controller (NNC), as shown in figure 8. The NNM sub-network acts as the dynamic model of the plant, providing a channel through which the errors could be backpropagated to adjust the parameters of the NNC adaptive controller. The two sub-networks operate with different time schedules during a sample interval. First the NNM is employed, to produce a predicted output track for the nonlinear plant. The errors, e_s, produced between the NNM predicted track and the actual plant output track are then used, following equation 13, to produce a cost J_m. The weights of this NNM are therefore updated as follows:

$$W_m(k) = W_m(k-1) - \frac{\lambda_m}{T_m} \sum_{i=k-T_m+1}^{k} \nabla J_m(W_m, i) \quad (18)$$

Here $W_m(k)$ represents the weight vector of the NNM sub-network and $\nabla J_m(W_m,k)$ is the gradient vector.

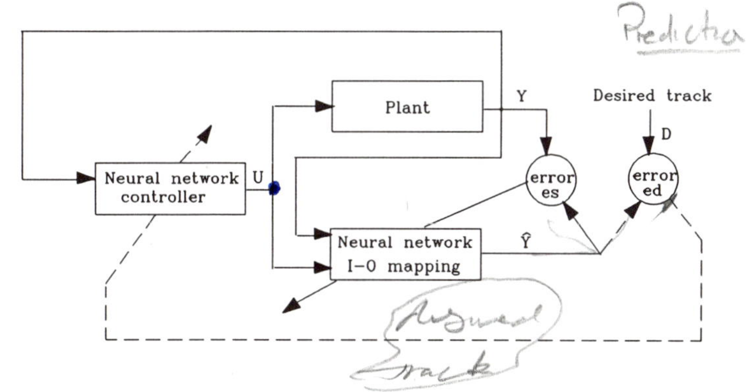

Figure 8 Neural regulator for adaptive control

Next, following equation 17, the cost J_c is produced from the errors, e_d, between the desired system output tracks and the resulting output tracks for the NNM, as shown in figure 8. The weights of the NNC sub-network are then updated following:

$$W_c(k) = W_c(k-1) - \frac{\lambda_c}{T_c} \sum_{i=k-T_c+1}^{k} \nabla J_c(W_c\ ,\ i) \tag{19}$$

Here $W_c(t)$ represents the weight vector of the NNC sub-network and $\nabla J_c(W_c,k)$ is the gradient vector. At this stage the NNM functions as a channel for the error backpropagation and its weights are unchanged. The outputs of the control sub-network are then sent to control the plant and are also taken as inputs to the NNM for the next mapping stage.

4.4.3 Implementation of the neural network regulator

Figure 9 illustrates the NN regulator for turbogenerator control. The orders p, q and s, r in equations 12 and 16 are all chosen to be 2, based on previous experience with the identification of ARMA models for turbogenerators, [14]. The hidden layer elements of both the NNC and the NNM sub-networks, contained the same nonlinear activation function as before. The two sub-networks were connected in cascade, with the input and output elements of both sub-networks chosen to have simple linear activation functions. The (6, 8, 2, 12, 14, 2) neural network regulator was implemented on the simulated model of the turbogenerator. The weights in the NNC were updated by the errors between the outputs

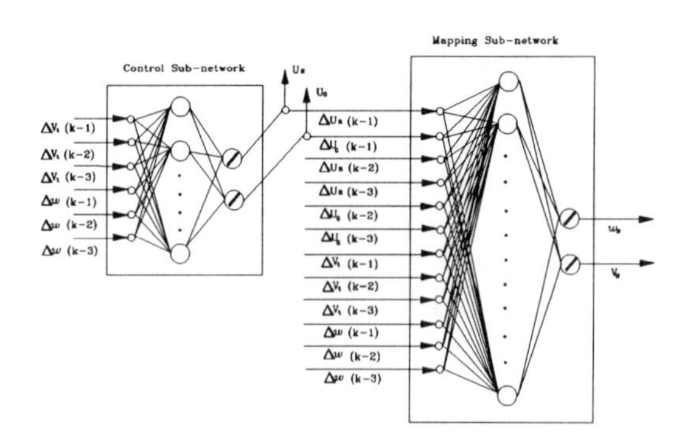

Figure 9 Hierarchial architecture for the NN regulator

of the NNM and the desired track, provided by a predictor which had been designed previously in [14]. This neural network regulator operates online as follows:

1. Y(k) and D(k) are sampled, and the vectors of time series shifted.
2. The patterns {Y(k-1)} and {U(k-1)} are input to the neural network regulator.
3. The weights in the NNM are updated to minimize the J_m.
4. The patterns {Y(k-1)} and {U(k-1)} are input again to the NN regulator.
5. The weights in the NNC are updated to minimize the J_c.
6. The control input vector U(k) is applied to the turbogenerator.

The step length λ_c used in training the NNC was set at 0.1, with a step length λ_m of 0.2 used in training the NNM subnetwork. T_c was chosen to be 1, with T_m set again at 10 samples.

With the initial operating conditions for the real and reactive powers being P_t = 0.8 p.u. and Q_t = 0.2 p.u, respectively, the regulator was trained using the PRBS signals discussed previously. After 100 seconds of training the performance of the neural network regulator was evaluated by introducing changes of setpoints and also by simulating large disturbances caused by short-circuits on the transmission lines. Figure 10(a) shows the response to changes in the terminal voltage setpoint of ±5% at 3 second intervals, in the presence of additive random noise on the terminal voltage. The response is well damped and follows the demanded changes without offsets in terminal voltage or

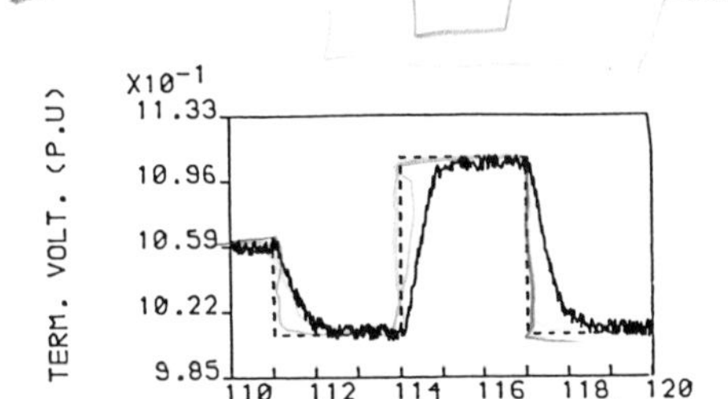

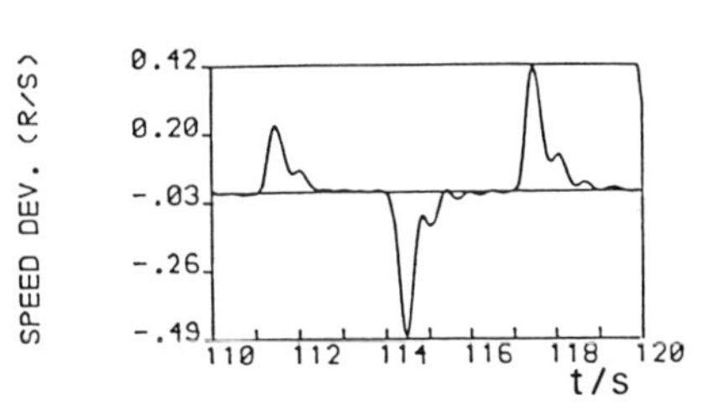

a) Voltage and speed dev. responses for setpoint changes

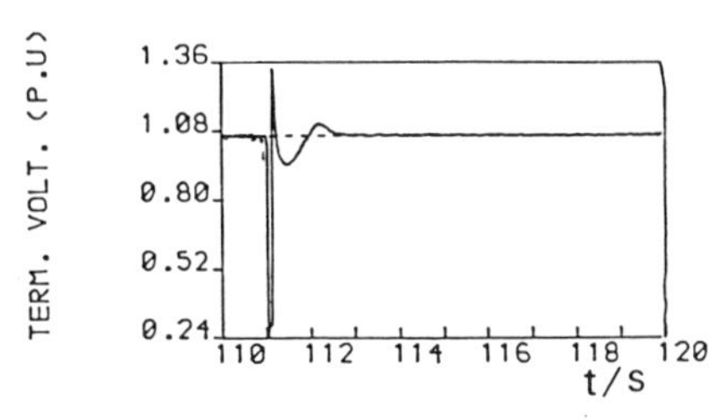

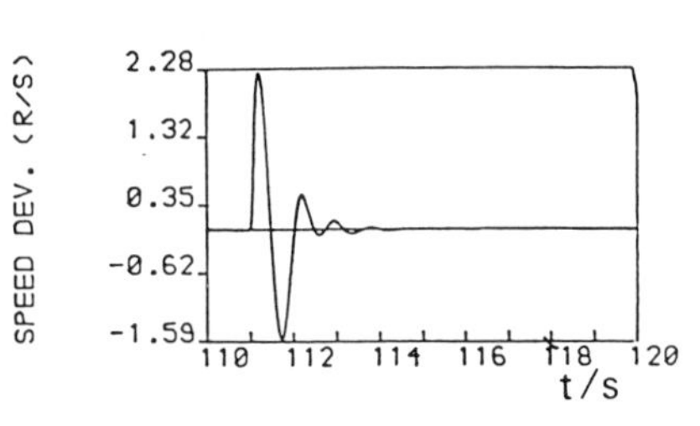

b) Voltage and speed dev. responses for S/C fault test

Figure 10 Test responses for initial conditions P_t=0.8p.u. Q_t=0.2p.u.

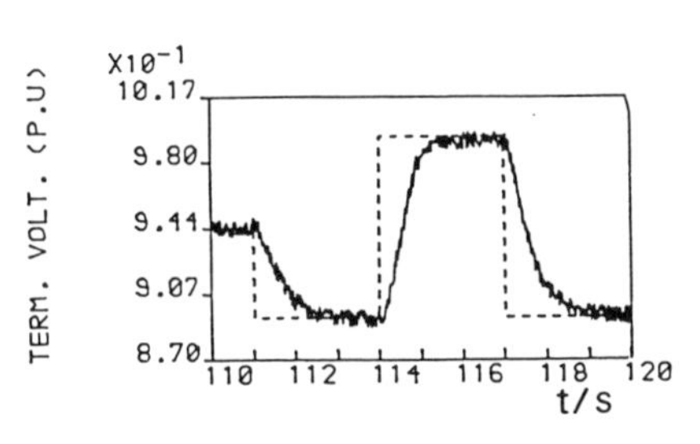

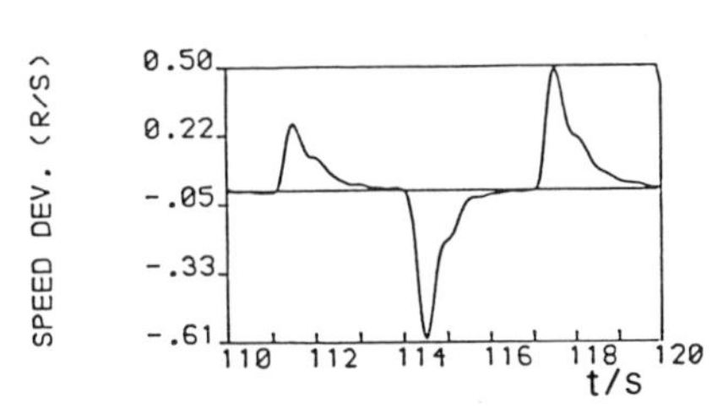

a) Voltage and speed dev. responses for setpoint changes

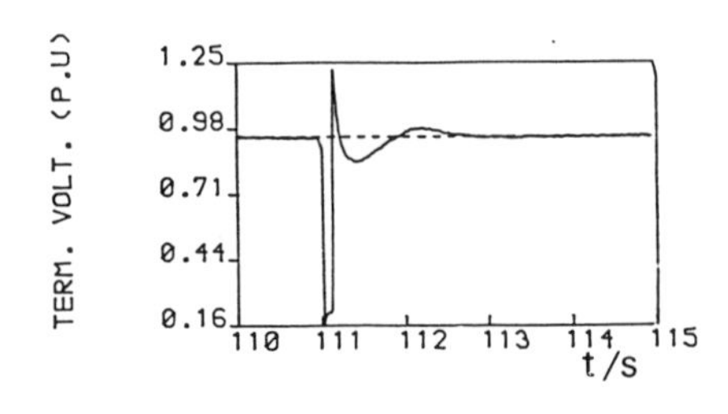

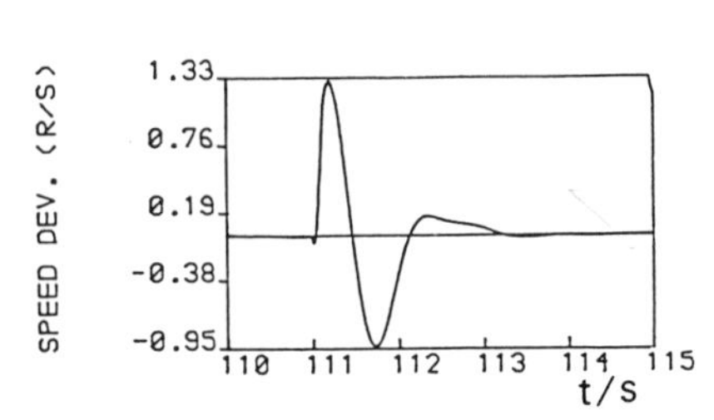

b) Voltage and speed dev. responses for S/C fault test

Figure 11 Test responses for initial conditions P_t=0.5p.u. Q_t=-0.1p.u.

speed. Figure 10(b) shows the response to a three-phase short circuit to earth at the sending end of one major transient disturbance. This fault causes the system to operate in a highly nonlinear mode, but again these responses are well damped.

The neural network regulator was also evaluated at other operating points. Figure 11 shows the controlled system performance with the initial operation condition $P_t = 0.5$ p.u. and $Q_t = -0.1$ p.u. In this case the turbogenerator system is operating with a leading phase and the performance is satisfactory. From figures 11a and 11b, it is seen that this regulation scheme can provide good control performance under different conditions. This can be ascribed to its nonlinear structure which has the flexibility to produce arbitrary complex relationships between plant outputs and control action as compared with existing adaptive controllers, [14], in which a linear combination of outputs is required to improve system damping.

4.5 Conclusions

This chapter introduced the concept of direct, neural, adaptive control, based on the ability of the Multilayer Perceptron to form a complex nonlinear adaptive control law. It was recognised that the position of the neural network, within this proposed controller structure, would yield problems with adjusting the network weights to achieve the desired system performance. A number of possible training techniques were introduced to overcome this problem. To allow for the use of the backpropagation training rule, it was proposed that either an approximation to the plant Jacobian could be used to determine the necessary gradients of the cost at the network output. Likewise a parallel neural model of the nonlinear plant could be trained and then used either as a channel for error backpropagation, or from analysis this could form an analytical expression for the plant Jacobian. Alternatively, it was proposed that search based optimisation techniques could be used, as these are not gradient based methods.

A structure for a continuous, direct, neural network-based, adaptive controller, was introduced. This utilised a nonlinear adaptive neural controller in parallel with a fixed gain linear control law. The neural control scheme was compared with traditional Lyapunov MRAC for the control of two simple nonlinear plants. It was found that, though the neural control method took longer to converge than the Lyapunov technique for the near-linear plant, it could provide accurate control for the highly nonlinear plant, for which the Lyapunov controller would not converge. Also, once trained, the weights of the neural network controller could be set, to form a fixed-gain, nonlinear control law, that would control the nonlinear system to follow a desired linear reference model.

The chapter finally described a novel approach to on-line dynamic modelling and adaptive control of turbogenerators. This employed a hierarchical architecture of neural networks which avoided the need to approximate the plant Jacobian, during the back propagation procedure. The neural controller structure did not require the use of a reference model or an inverse model of the plant. The neural network regulator had a compact structure, which could easily be extended to cater for more complex dynamic systems or additional control loops. The performance of this regulator was evaluated by simulation on a detailed nonlinear model of a turbogenerator system. It was found that the neural regulator provided accurate control, with good damping of transients, for disturbances occurring at various operating conditions.

Acknowledgements

The authors would like to acknowledge the support of the Institute of Advanced Microelectronics and Dupont (UK) Ltd., Londonderry.

References

[1] Astrom, K.J. and Wittenmark, B., "Adaptive control", Addison-Wesley Publishing Company, Reading, Mass., 1989.

[2] Cybenko, G.,"Approximations by superpositions of a sigmoidal function", Mathematics of Signals and Systems, Vol. 2, pp. 303-314, 1989.

[3] Funahashi, K., "On the approximate realisation of continuous mappings by neural networks", Neural Networks, Vol. 2, pp. 183-192, 1989.

[4] Banks, S.P., "Control System Engineering", Prentice-Hall International, London, U.K., 1986.

[5] Rumelhart, D.E., Hinton, G.E., and Williams, R.J., "Learning internal representations by error propagation", in Rumelhart, D.E., and McClelland, J.L., (Eds), Parallel Distributed Processing, Vol.1, MIT Press, 1986.

[6] Psaltis, D., Sideris, A. and Yamamura, A., "Neural controllers", Proc. of 1st International Conference on Neural Networks, Vol.4, pp.551-558, San Diego, USA., 1987.

[7] Saerens, M. and Soquet, A., "A neural controller", Proc. 1st IEE Conference on Neural Networks, pp.211-215.

[8] Lightbody, G. and Irwin, G.W., "Neural networks for nonlinear adaptive control", Proc. IFAC Symposium on Algorithms and Architectures for Real-Time Control, Bangor, U.K., 1991.

[9] Chen, S., Billings, S.A. and Grant, P.A. "Non-linear system identification using neural networks", Int. J. Control, Vol. 51, No. 6, pp. 1191-1214, 1990.

[10] Nguyen, D., and Widrow, B., "The truck backer-upper: and example of self-learning in neural networks", Proc. International Joint Conference on Neural Networks, Washington, D.C., Vol.2, pp.11357-11363, June, 1989.

[11] Jordan, M.I. and Jacobs, R.A., "Learning to control an unstable system with forward modelling", Advances in Neural Information Processing Systems, Vol. 2, pp. 324-331, Morgan Kaufman, San Mateo, CA., 1990.

[12] Hunt, K.J. and Sbarbaro, D., "Neural networks for nonlinear Internal Model Control", Glasgow University, Dept. of Mech. Eng., Internal Report, 1990.

[13] Bleurer, H., Diez, D., Lauber, G., Meuer, U. and Zlatnik, D., "Nonlinear neural network control with application example", Proc. Int. Neural Net. Conference, INNC 90, Vol. 1, pp. 201-204, Paris 1990.

[14] Wu, Q.H., and Hogg, B.W., "Adaptive controller for a turbogenerator system", IEE Proc., Vol.135, Pt.D, No.1, 1988, pp.35-42.

[15] Wu, Q.H., and Hogg, B.W., "Robust self-tuning regulator for a synchronous generator", IEE Proc. Vol.135, Pt.D, No.1, 1988, pp.463-473.

[16] Narendra, K.S., and Parthasarathy, K., "Identification and control of dynamical systems using neural networks", IEEE Trans. on Neural Networks, Vol.1, No.1, 1990, pp.4-27.

[17] Bassi, D.F., and Bekey, G.A., "High precision position control by cartesian trajectory feedback and connectionist inverse dynamics feedforward", Proc. International Joint Conference on Neural Networks, Washington, D.C., Vol.2, pp.11325-11331, June, 1989.

[18] Chen, F.C., "Back-propagation neural network for nonlinear self tuning adaptive control", Proc. of IEEE international Symposium on Intelligent Control, 25-26 September, 1989, Albany, New York, pp.274-279.

[19] Wu, Q.H., Hogg, B.W., and Irwin, G.W., " A neural network regulator for turbogenerators", IEEE Trans. on Neural Networks, Vol. 3, No. 1, 1992, pp. 95-100 .

Chapter 5

Comparative aspects of neural networks and fuzzy logic for real-time control

C. J. Harris

5.1 INTRODUCTION

The structure and consequent outputs in response to external commands for intelligent controllers is determined by experiential evidence or the observed input/output behaviour of the plant, rather than by reference to a mathematical or model based description of the plant. Intelligent control is an amalgam of the disciplines of artificial intelligence, control theory and operational analysis, requiring few limitations upon the linearity, time invariance, deterministic or non-complex nature of the plant. It is the authors view that there is little to be gained by applying intelligent control to linear time invariant systems, a wealth of model based classical control theory suffices for such systems. Intelligent control is properly aimed at processes that are ill-defined, complex, nonlinear, time varying and stochastic. Two approaches dominate the real time (that is online) intelligent control field: (i) Neural Net Controllers (Neurocontrollers) (Brown and Harris, 1991), (ii) Self Organising Fuzzy Logic Intelligent controllers (SOFLIC's) (Harris et al, 1992). In the following we will consider both these classes of controller and their relevence to real time control and guidance.

Fuzzy logic controllers maybe either static systems, which have fixed production type rule bases, or adaptive or self organising that rewrite the rule base by experiential means. *Direct* self-organising fuzzy logic controllers use observations of the system closed loop performance to directly manipulate the controller fuzzy rule base or relational matrix, without any intermediate process model being produced, (Sutton, 1991). Whereas the *indirect* SOFLIC, (Moore and Harris, 1992), generates a fuzzy relational matrix of the plant and then inverts this model in order to find the control which realises the desired next state. The indirect method has the advantage that the system's response can be changed by simply supplying a new desired next state rule base, rather than retraining the whole fuzzy controller. Benchmark IFAC examples, (Moore, 1991), indicate that the indirect SOFLIC is extremely robust with respect to parametric variations, external disturbances, system nonlinearities and system failures. The separation of the adaptation system from the controller design enables rapid learning convergence to be established, (Brown and Harris, 1991). SOFLIC's are considered in depth in section 5.2. For more details including benchmark/design examples see Harris et al (1992).

Artificial Neural Networks (ANN) are much older than SOFLIC's (although not in application) originating from work in the 1950's. ANN's have many desirable properties that make them suitable for intelligent control applications.

1. ANN learn by experience rather than by modelling or programming.

2. ANN have the ability to generalise, that is map similar inputs to similar outputs.

3. ANN can form arbitrary continuous nonlinear mappings.

4. ANN architectures are distributed, inherently parallel and potentially real time.

   Additional ANN properties are required for neurocontrollers.

5. Temporal stability, the ability to absorb new information (plasticity) whilst retaining knowledge (or rules) previously encoded across the network (stability).

6. Real time adaptation or learning in response to plant variations.

7. Known or proven learning convergence conditions necessary for process closed loop behaviour prediction or controller certification.

The most popular ANN's in neurocontrol are the Multi-Layer Perception (MLP), (Narendra, 1990), and the Backward Error Propagation (BEP) algorithm. Unfortunately MLP's are slow, convergence cannot be established, and increasing the state space covered by the training set results in the whole network being retrained. Additionally MLP's are temporally unstable (Brown, 1990). The poor convergence rates of the MLP have been overcome by essentially preprocessing in the Functional layer Link Network (FLN), (Pao, 1989); however the preprocessing requires knowledge of the family of nonlinear trasformations carried out by the plant. Associative memory single layer networks such as Albus CMAC (see also Chapter 8), Radial Basis

Function (RBF) (see Chapter 9) and the Brown B-Spline Network (Brown and Harris, 1992), all satisfy in some measure conditions 1-7 above. Since Associative Memory ANN's first form a nonlinear transformation of the input space (usually to a much higher dimensional space), followed by a single *linear* layer, which forms a weighted combination of the transformed input. The first layer is generally fixed, in which some apriori knowledge about the plant can be incorporated, whereas the output layer contains weights that are adjusted by output error feedback - resulting in a single convex cost function in the error-weight space for which there exists a single global minima. Unlike in the MLP in which nonlinear optimisation is carried out over a complex cost function with many local minima which traps gradient descent rules. Equally, the SOFLIC's are fixed nonlinear input mappings into a higher dimensional space, followed by a defuzzification process (such as centre of area), which linearly combines the outputs of the input layer in proportion to their respect confidence or belief. If these confidences can be adjusted by a learning rule, by say output error feedback, then the input/output process is self-organising.

In this Chapter, aspects of single layer associative memory neural nets are considered for neurocontrol and guidance applications, it also is shown that there is direct linkage between this class of ANN and Self-Organising Fuzzy Logic Systems if centre of area defuzzification is used together with product operators.

5.2 SELF-ORGANISING FUZZY LOGIC INTELLIGENT CONTROL (SOFLIC)

5.2.1 Introduction

Fuzzy logic control attempts to solve complex control problems by relaxing the specification of controlled response in manner that reflects the uncertainty or complexity of the process under control. For example, car driving is complex to mathematically model, and greatly influenced by temporal unkown a priori environmental conditions (road surface, weather etc). Yet adequate control can be achieved by humans, utilising only experiential models and associated learning. Human control performance may not be precise (as compared with an algebraic controller), but the goals to be achieved and tolerances required are rather vague. The human control 'algorithm', effectively rejects disturbances, copes with nonlinearities, and learns or adapts to new situations. This performance is achieved through a trade off between precision and significance. An internal model of driving is being used, but is somewhat vague, perhaps utilising a large number of vague or fuzzy situation dependent rules, which are combined by a means of rule interpolation for a given situation to yield a 'best' situation prediction.

Fuzzy Logic (Zadeh, 1963) was introduced to model human reasoning by giving definitions to vague terms and allowing several rules in a rule base to interact with varying degrees of belief. It is irrelevent whether or not humans use this mechanism, what is important is that fuzzy logic enables an interacting rule base with vague terms with the property of generalisation to be generated. The purpose of SOFLIC's, with the inherent properties of generalisation (associative memory) and rule interaction is to produce controllers, that require little a priori plant knowledge, can adapt to temporal plant changes and are robust to nonlinearities and disturbances.

5.2.2 Fuzzy logic controllers

Zadeh first suggested fuzzy logic for control in 1972, however the most significant work was carried out by E.H. Mamdani and his students at Queen Mary College during the 1970's (Mamdani, 1974, Mamdani and Assilan, 1975). More recently static fuzzy logic controllers have become very popular in Japan through their natural rapid prototyping, simplicity of implementation and reduced design complexity. For static fuzzy logic controllers the rule base is usually produced by eliciting knowledge from human operators (cf. Expert Systems) - somewhat limiting their applicability.

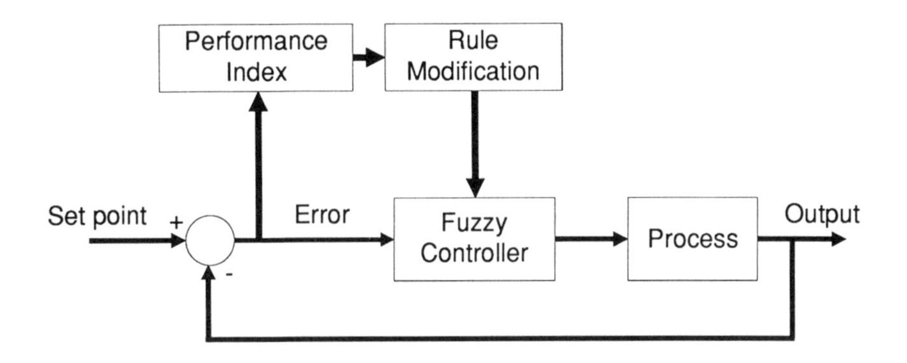

Figure 5.1 Self-organising fuzzy controller structure

Procyk and Mamdani (1979), introduced the *direct* adaptive fuzzy controller - see figure 5.1. In which a fixed performance table is used to determine modifcation to the adaptive control rule base, without the intermediate step of process identification. This direct self-organising controller (DSOC) has obvious advantages over the static fuzzy logic controller, in the amount of a priori knowledge required and the ability to respond to plant changes. Examples and design procedures can be found in Harris et al (1992).

It is important to note that both the static and direct adaptive fuzzy logic controllers operate on error signals only. Assuming that the controlled response is independent of plant output state, dependent only on error. If a consistent response is to achieved over the full operating envelope, the process response, to inputs must be independent of current system output - this may be true of linear plants, but can not be so for nonlinear processes. A major reason behind the indirect adaptive fuzzy controller (or SOFLIC).

5.2.3 Indirect adaptive fuzzy controllers

Indirect adaptive fuzzy controllers or SOFLIC's, separate model identification and control, allowing closed loop response to be modified without causing any change in plant model. The SOFLIC structure is shown in figure 5.2; the controller rule base contains the fuzzification, composition and defuzzification of a conventional fuzzy logic controller, but its rules are based upon the actual process outputs rather than error signals. The control rules are computed on-line via the inverse of the fuzzy model. This approach offers separate analysis of long term stability (model adaptation) and short term stability (closed loop stability), ability to change controller performance requirements *online*, and the use of model data to infer process health/performance.

5.2.4 Fuzzy models

In the following it is assumed that the basics of fuzzy logic are known to the reader (if not see Harris 1992). For a variable $y \varepsilon Y$ (a universe of objects or discourse), and a subset $A \subset Y$, $y \varepsilon A$ has membership value $\mu_A(y)$ ($\mu_A : Y \rightarrow [0, 1]$). To construct a fuzzy set model, mappings from one universe of discourse to another (i.e. input $\rightarrow$ output mappings) can be achieved by fuzzy implications or rules:

$$\text{IF } A(u) \text{ THEN } B(y)$$

which links condition set A (defined by $\mu_A(u)$, $u \varepsilon U$) with an output set B (defined by $\mu_A(y)$, $y \varepsilon Y$). In general there are N such rules to define a mapping,

	IF $A_1(u)$	THEN	$B_1(y)$
OR	IF $A_2(u)$	THEN	$B_2(y)$
.	.	.	.
.	.	.	.
.	.	.	.
OR	IF $A_N(u)$	THEN	$B_N(y)$

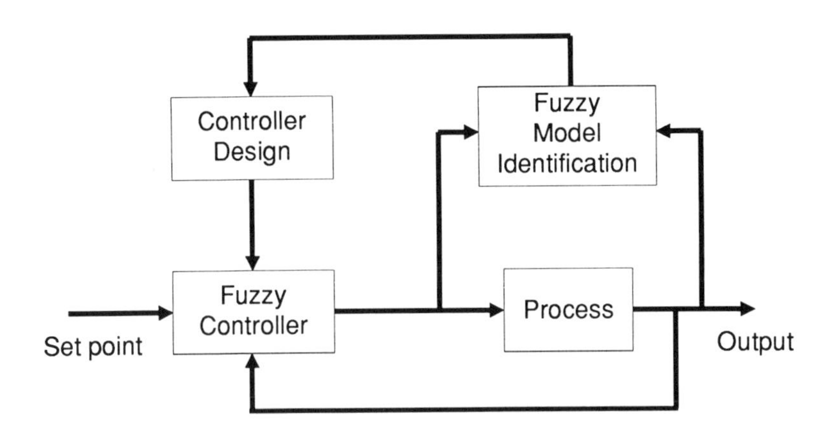

Figure 5.2. Indirect adaptive fuzzy controller

A fuzzy relation, R, is a function of two variables such that its membership function, $\mu_R(u, y)$ corresponds to the strength of connection between u and y in the mapping. The fuzzy relation for the ith rule, R_i, is given by the intersection of the condition and output sets

$$R_i(u, y) = A_i(u) \cap B_i(y) \tag{5.2.1}$$

A complete algorithm, R, is given by the union of the relation formed for each rule,

$$R(u, y) = \bigcup_i R_i(u, y) \tag{5.2.2}$$

For SOFLIC's, the number of rules, N is unkown, and R is time varying ($R(t)$). It is assumed that every observation pair $\{A_t(u), B_t(y)\}$ corresponds to a new rule to be added to $R(t)$ to create a new relation $R(t + T)$ (T = sample period) : i.e.

$$R(t+T) = R(t) \cup \{A_t(u) \cap B_t(y)\} \tag{5.2.3}$$

This is adequate if new observations are consistent with exisiting encoded rules, however for time varying processes, new observations must replace old rules when inconsistencies occur through a forgetting factor D. That is the existing relation R is decayed to an intermediate relation $R'(t)$ through

$$\mu_{R'(t)} = \mu_{R(t)} \times D \tag{5.2.4}$$

Which is used to form the new relation R (t+T) through:

$$R(t+T) = R'(t) \cup \{A_t(u) \cap B_t(y)\} \tag{5.2.5}$$

For dynamical systems, state derivatives need to be incorporated into the algorithms, for example for a first order fuzzy model

$$Y \times U \xrightarrow{R} \dot{Y} \tag{5.2.6}$$

maps the input universe U and output universe Y onto the output derivative $\dot{Y}$. The associated fuzzy relation, R, is a collection implication statements of the form:-

$$R_i : IF\ y_i(t)\ AND\ u_i(t)\ THEN\ \dot{y}_i(t+T_d) \tag{5.2.7}$$

where T_d is a delay parameter representing process dead time, and is equivalent to the delay in reward parameter used in DSOC algorithms.

5.2.5 SOFLIC design

It is assumed that the above process has generated an adequate fuzzy model of a plant G, through the adaptive relational matrix R. Then, in a manner similar to the certainty equivalence principle, an online controller can be derived from the concept of causality inversion of the causal fuzzy relation R, defined by equations (5.2.6) and (5.2.7). The problem is now, given the required output derivative, $\dot{y}_i(t+T)$ at some time T in the future what input $u_i(t)$ has to be applied to the plant G to achieve it? This inverse causality mapping can be described by a fuzzy relation, R^{-1}, such that

$$Y \times \dot{Y} \xrightarrow{R^{-1}} U \tag{5.2.8}$$

which is made up of a collection of rules of the form:

$$R_i^{-1} : IF\ y_i(t)\ AND\ \dot{y}_i(t+T)\ THEN\ u_i(t) \tag{5.2.9}$$

The problem here, is given a relational matrix R, can the corresponding R^{-1} be evaluated? There are basically two methods:

(i) Causality inversion via associative memory.

Consider a particular observation $\{y_i(t), u_i(t), \dot{y}_i(t + T)\}$, using the rule structure above for R, the membership values for this rule are given as in equation (5.2.1) by

$$R_i(Y, U, \dot{Y}) = y_i(t)(Y) \cap u_i(t)(U) \cap \dot{y}_i(t + T)(\dot{Y}) \tag{5.2.10}$$

If this observation is also considered to be an element of R^{-1} then the membership value of this inverse rule can also be evaluated. Using the structure for R^{-1} shown above, then the membership values are given by:

$$R_i^{-1}(Y, U, \dot{Y}) = y_i(t)(Y) \cap \dot{y}_i(t + T)(\dot{Y}) \cap u_i(t)(U) \tag{5.2.11}$$

Comparing (5.2.10) with (5.2.11), the only difference is in the order in which the individual sets occur. Via the commutative property of the intersection operator, it is clear that these two relation elements have identical values, i.e. it is possible to perform the inverse causality mapping using the compostion operator, $\circ$, on the original relational matrix, R, simply by accessing the relation elements in the proper order.

(ii) Causality inversion via prediction

If the relational matrix R is used as a predictor, via the composition operator, $\circ$, for $y_i(t + T)$ for several values of $u_i(t)$ - a prediction from every possible rule is obtained by centring these u_i's on the discretisation points - then its causal inverse is given by that $u(t)$ for which the $y(t + T)$ prediction intersects the desired value for this variable. Moore (1991) has generated a small rule set for achieving this causality inverse. The technique works particularly well for partial rule situations, this is significant on initialisation of the SOFLIC as there are few (if any) rules available. The extrapolation technique/rule in this method allows rapid learning through the controller entering operating regions for which there is no previous experience.

Consider a plant with fuzzy relation G, with measured/observed input/output (u, y), which can be used to predict a new output, $y(t + T)$, using the compostion operator $\circ$,

$$y(t + T) = G \circ \{u(t) \times y(t)\} \tag{5.2.12}$$

A command signal, $d(t)$, is available together with the desired response between $d(t)$ and $y(t)$ through the performance relational matrix P:

$$y(t + T) = P \circ \{d(t) \times y(t)\} \tag{5.2.13}$$

The control design task is to find a feedback controller, C, that matches the closed loop response to the desired response. The fuzzy relation C, maps the demand $d(t)$ and output $y(t)$ to the control action $u(t)$ via composition:

$$u(t) = C \circ \{d(t) \times y(t)\} \tag{5.2.14}$$

Substituting (5.2.14) into (5.2.12) gives the closed loop response

$$y(t+T) = G \circ \{y(t) \times C \circ \{d(t) \times y(t)\}\} \tag{5.2.15}$$

Comparing this closed loop response with the desired response (5.2.13), provides the relationship

$$P \circ \{d(t) \times y(t)\} = G \circ \{y(t) \times C \circ \{d(t) \times y(t)\}\} \tag{5.2.16}$$

Provided that G^{-1} exisits, then the controller is given by

$$C \circ \{d(t) \times y(t)\} = G^{-1} \circ \{y(t) \times P \circ \{d(t) \times y(t)\}\} \tag{5.2.17}$$

A graphical representation of this is given in figure 5.3. The performance rule matrix, P, could be an algebraic or any other performance measure, all that is required is that its output is fuzzified before being passed to inverse process model G^{-1} (or into equation (5.2.17)).

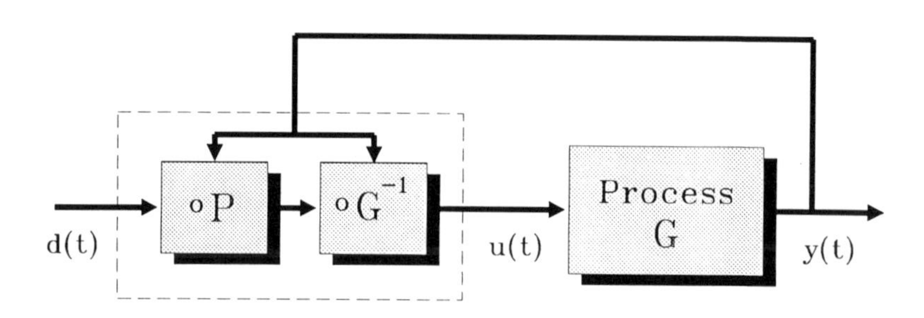

Figure 5.3. SOFLIC internal structure

5.3 NEUROCONTROLLERS

5.3.1 Introduction to ANN's in control

ANN are typically constructed from parallel layers of simple computational nodes, with weighting elements between nodes that define their stength of connection weights, which can be adapted during learning by some optimisation procedure to model some input/output map. ANN's may be categorised in a variety of ways, for the purposes of neurocontrol a differentiation between *local* and *global* generalisation is useful. Network generalisation is global if one or more of the networks weights can effect the network output for any and every point in the input space. MLP's are examples of globally generalising networks, this property leads to network robustness and generalisation, but at the expense of slow learning and network wide learning interference. Whereas local generalisation occurs in networks for which only a few weights effect the network output response for point within a local region of the input space. RBF and B-spline networks are examples of local generalisation networks, in which learning inteference is minimised and learning relatively fast, owing to the minimum number of weights to be updated.

Before considering the structure of various AAN's for neurocontrol, it is important to first establish their approximation capabilities to some arbitrary nonlinear continuous mapping $y = f(x) : D \subset R^n \rightarrow R^p$ from input/output data sets $\{x, y\}$ to which arbitrary accuracy, $f(x, \underline{w})$ on D, where D is a compact subset on R. Where $\underline{w}$ is the parameter weight vector that is updated by some learning rules (such as LMS, NLMS, etc.). The Stone-Weierstrass theorem (Harris, 1992) shows that for any continous function $f \varepsilon C(D)$, a compact metric space, an $\hat{f}$ can be found such that $|f - \hat{f}| < \varepsilon$ for an arbitrary $\varepsilon > 0$. Or by a corollary, for measurable functions f, any infinitely large network of elementary modules or computational elements can model any function in all but a few isolated points. To apply this theorem to ANN we need to restrict the class of ANN's to those encountered in neurocontrol. That is to those which are:-

(i) Adaptively *linear* in the input space to hidden layer, but are fixed nonlinearly in the hidden layer to output (e.g. MLP, FLN's). Leading to nonlinear optimisation in BEP is utilised. This class can be represented by

$$ANN_1 \{f \varepsilon C(D) / \hat{f}(\underline{x}) = \sum_{j=1}^{m} a_j \varnothing (\sum_{k}^{n} x_k w_{kj} + w_{jo}) , \quad w_{kj}, a_j, w_{jo} \varepsilon R\} \tag{5.3.1}$$

where a_j are constants, $\underline{w} = \{w_{kj}\}$ is the m-dimensional weight vector, and $\varnothing$ (.) is an invertible, differential squashing function such as a sigmoid.

(ii) Those networks which have *fixed nonlinear* input space to hidden layers, but have *adaptive linear* hidden layer to network output. Leading to linear parameteric optimisation (e.g. RBF's, B-splines, CMAC). This class can be represented by

$$ANN_2 \{f \varepsilon C(D) / \hat{f}(\underline{x}) = \sum_{j=1}^{m} a_j \psi_j (\| \underline{x} - \underline{c}_j \|) , \quad w_j \varepsilon R, \underline{c}_j \varepsilon R^n\} \tag{5.3.2}$$

where $\underline{c}_j$ are fixed centroids of the basis functions ψ_j (.) and w_j are adjustable weights.

It has been shown (Harris, 1992, Cotter, 1991) that both classes ANN_1/ANN_2 satisfy the Stone-Weierstrass theorem, but only class ANN_2 have unique existence sets (Chebychev sets). In addition, class ANN_2 provide unique optimal approximations, (i.e. there exists a single global minimum in the cost-weight space) if ψ_j (.) have compact support on D, or are equivalently locally piecewise polynominals. This latter property also ensures temporal stability in learning. It would appear that only class ANN_2 satisfy all the conditions 1-7 of section 5.1 required for effective neurocontrollers. There is a penalty to be paid in implementing neurocontrollers of class ANN_2, the fixed nonlinear input mapping increases the dimension of the input space, practically limiting their applicability to low dimension systems. Chapters 8 and 9 consider the CMAC and RBF forms of class ANN_2, and will not be considered further here, although detailed studies can be found in Chen & Billings (1992) or Harris et al (1992).

5.3.2 The B-Spline neurocontroller

The B-spline neural network (BSNN) was developed for use in on-line adaptive modelling and control, as well as for static off-line design such as guidance laws, trajectory planners etc. A BSNN is of class ANN_2, an is constructed from a linear combination of basis functions ψ_j (.) $\equiv a_j \varepsilon C(D)$ (which is also strictly convex as well as compact) which are piecewise polynominals of order k. Figure 5.4 illustrates basis functions of order 1, 2, 3;

note the similarity to fuzzy sets, and their compact support. The output of the BSNN is

$$y = \underline{a}^T \underline{w} \tag{5.3.3}$$

where $\underline{a}$ (ε $[0, 1]^p$) is the bais function output vector, and $\underline{w}$ a weight vector. (In the following only scalar outputs are considered without loss of generality). The network output is a piecewise polynominal of order k (selected by the designer) that provides a unique best approximation to a nonlinear function f ($\underline{x}$) (x ε R^n). the basis functions, $\underline{a}$, nonlinearly map the input space $\underline{x}$, into a higher dimensional space in which the desired mapping is approximately linear. The basis functions have finite support, so that only a small number $\rho = k^n$ of the variables, a_i, are non-zero from the transformation $\underline{x} \longmapsto \underline{a}$. For adaptive modelling and control this map is fixed, adaptation only occurs through backward error propagation to the linear set of weights $\underline{w} = \{w_i\}$ ($i = 1, \ldots, p$). any first order weight updating or learning laws such as Least Mean Squares (LMS), or Stochastic Approximation LMS may be used (see section 5.3.3) to ensure convergence in approximation. These learning laws are real time owing to the fact that only a few variables, a_i, associated with a particular input, are nonzero, and are temporally stable.

For off-line or static design, the compact support nature of BSNN offers transperancy between network output and a particular weight. And like fuzzy logic, the BSNN can be used for static design aspects whilst satisfying a set of constraints. Since the local compact support nature is exploited to include boundary conditions that only modify network computations locally.

The polynomial basis functions used in the BSNN are splines (Cox, 1990), which provide piecewise polynomial approximations; instead of approximating an arbitary function by a series of polynomials over an interval the function is approximated by polynomials defined over *subintervals* of the partition space. By which *interpolation* between *knots* or partition nodes of the complete interval is used to represent the function $f(\underline{x})$. Univariate and multivariate B-splines may be readily found by simple recursive equations (Harris, 1992); a nice property of B-Splines is that simple recursive equations for their derivative and integral may also be found - that enable dynamic process constraints to be readily incorporated into the approximation.

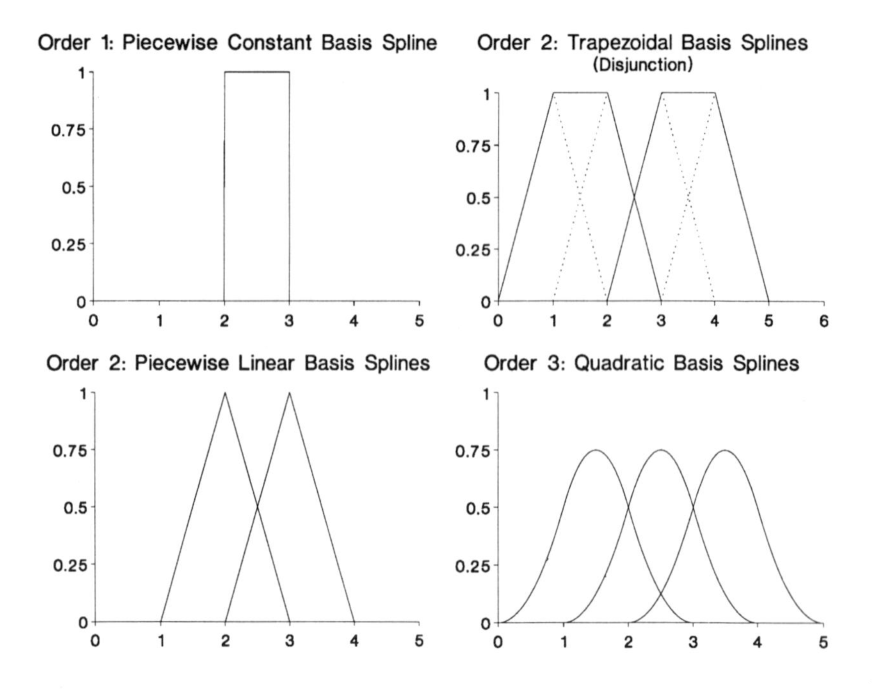

Figure 5.4. B-Spline functions, order 1-3

5.3.3 BSNN weight updating

Given a set of observed input/output data pairs $\{y(r), x(r)\}^{M}_{r=1}$ (where $M \geq p$; p the dimension of the weight vector); the optimal weight vector is given by the (pseudo) inverse of (5.3.3) or

$$\underline{A}\,\underline{w} = \underline{y} \tag{5.3.4}$$

where $\underline{y}$ is the output vector of observations of dimension $(M \times 1)$ and $\underline{A}$ is an $(M \times p)$ matrix, whose rth row corresponds to the basis function output for input $\underline{x}(r)$. Hence the optimal weight vector, $\underline{w}$, is given by

$$\underline{w} = (\underline{A}^T \underline{A})^{-1} \underline{A}^T \underline{y} \tag{5.3.5}$$

provided that $(\underline{A}^T\underline{A})$ is invertible. However for every new data pair, (5.3.3) has to be re-evaluated, with all its associated computational overheads. Recursive Least Squares (RLS) is possible, but has practical problems associated with local learning and computational overheads ($O(p^2)$ floating point operations). Whereas the LMS and Normalised LMS (NLMS) weight update rules use an instantaneous estimate of the current value of the mean squared error (MSE), providing convergence to the MSE only is the limit. Despite this the LMS algorithm only updates those weights whose basis function is non-zero requiring only $(\rho + 1)\, f$ floating point operations.

The LMS rule is given by

$$\Delta \underline{w}(n) = \delta (y(r) - \underline{a}^T(r)\, \underline{w}(r-1))\, \underline{a}(r) \tag{5.3.6}$$

for which the absolute value of the normalised error, e, reduction can be shown to be:

$$\left| \frac{\Delta e(r)}{e(r)} \right| = \delta\, \underline{a}^T(r)\, \underline{a}(r) \tag{5.3.7}$$

Hence for convergence of the LMS weight updating rule for the BSNN

$$0 < \delta < 2\, (\max_r \{\underline{a}^T(r)\, \underline{a}(r)\})^{-1} \tag{5.3.8}$$

Unfortunately the normalised approximation error reduction depends upon the variance of $\{\underline{a}^T(r)\, \underline{a}(r)\}$, which is in turn dependent upon the order, k, of the basis splines selected by the designer. This problem can be avoided by using the NLMS algorithm:-

$$\Delta \underline{w}(r) = \delta (y(r) - \underline{a}^T(r)\, \underline{w}(r-1)\, \frac{\underline{a}(r)}{(\underline{a}^T(r)\, \underline{a}(r))} \tag{5.3.9}$$

The use of the normalisation term in (5.3.9) means that the NLMS rule is an error correction rather than gradient descent rule. The NLMS weight update rule is self-scaling, and for convergence the updating factor $\delta\ \varepsilon\ (0, 2)$.

In the LMS/NLMS algorithms, the correction gain factor δ is constant, however by varying its magnitude during the learning process can provide fast initial convergence yet in the long term effective filtering of measurement and noise. This can be readily achieved by (i) assigning an individual learning rate to each basis function and (ii) reduce δ, as a function of iteration and as confidence in a particular weight increases. The Stochastic Approximations LMS/NLMS algorithms provide these attributes through:

$$SLMS\text{: } \Delta w_i(r) = \delta_i(r)\, (y(r) - \underline{a}^T(r)\, \underline{w}(r-1))\, a_i(r)$$

$$SNLMS\text{: } \Delta w_i(r) = \Delta_i(r)(y(r) - \underline{a}^T(r)\,\underline{w}(r-1)) \frac{a_i(r)}{\underline{a}^T(r)\,\underline{a}(r))} \tag{5.3.10}$$

where these rules converge in the MSE sense with probability one, if the correction factor, δ_i, satisfies the Robbins-Munro conditions:

$$\delta_i(r) > 0, \quad \lim_{r \to \infty} \delta_i(r) = 0$$

$$\sum_{r=1}^{\infty} \delta_i(r) = \infty, \quad \sum_{r=1}^{\infty} \delta_i^2(r) < \infty \tag{5.3.11}$$

It is worth noting that in the original CMAC, the training rule used is the same as (5.3.6), but with δ replaced by $\delta\rho^{-1}$, but as $\underline{a}^T(r)\,\underline{a}(r)$ can be shown equivalent to ρ, the CMAC weight update is equivalent to *both* LMS and NLMS training rules.

A significant result in the BSNN, is its local learning of new information, since when the desired function, f, is modified locally (through say parameter changes), the recommened weight changes by the LMS rule decay exponentially for weights far from this modified domain. This relaxes the requirement that the input signal be persistently exciting over the global domain, to that over a local domain.

The B-spline neural net designer has to choose the order, k, of the splines. Piecewise linear B-splines provide computationally cheap and adequate models for most practical situations, being able to learn quickly with very limited learning interference. Higher order models imply smoother output surfaces, but also give increased learning interference with associated slower learning rates. Simultaneous increase of B-spline order with reduction in the partitioned intervals, reduces learning interference (Brown, 1992).

5.4 COMPARATIVE ASPECTS OF FUZZY LOGIC AND ANN'S

In fuzzy logic (FL) there are many operators used for logical implication, it has been demonstrated (Brown and Harris, 1991) that the Larsen *product* operator provides superior interpolation properties over the more usually used *MIN* operator. (This is not surprising since the product operator retains more information in the output). Suppose that centre of area defuzzification is used, together with the product operator for implication, to generate the output, y_o, of a fuzzy model i.e.

$$y_o = \sum_1 \mu_{U_i}(u)\, w_i \Big/ \sum_i \mu_{U_i}(u) \tag{5.4.1}$$

where $w_i \triangleq \Sigma\, c_{ij}\, y_i$ is the weight associated with the input set U_i, and c_{ij} is the confidence that the output is y_i if the input set U_j has set membership of unity (equivalently c_{ij} is the ijth element of the fuzzy model relational matrix). If normalised B-splines are used to model the fuzzy sets U_i, and the product operator is used for logical *AND*, then $\Sigma\, \mu_{U_i} \equiv 1$, and (5.3.12) is simplified to

$$y_o = \sum_i \mu_{U_i}(u)\, w_i \tag{5.4.2}$$

That is the fuzzy sets, U_i, form a *basis* for the defuzzified output y_o, which is linearly dependent upon the fuzzy sets through the weights w_i. (5.3.13) is identical in structure to a single layer associative memory neural net such as (5.3.3) (or the RBF, CMAC etc). A fuzzy logic controller can be envisaged as an input ⇒ input fuzzy set map (a fixed nonlinear transformation), then an input fuzzy set ⇒ output map via the defuzzification rule (5.3.13). The second mapping is equivalent to a single layer neural net, which maybe adaptive if backward error propagation is used to change the weights w_i. In this case, the fuzzy logic controller becomes a SOFLIC, and if SLMS or SNLMS weight update algorithms are used, the convergence properties of single layer associative memory ANN apply equally to SOFLIC's. Whilst the relationship between SOFLIC's and ANN's now appears obvious, little theory exists in the literature for establishing the learning convergence of SOFLIC's. Brown and Harris, (1991) have derived a family of weight updating rules for SOFLIC's that exploit this relationship, such that real time on-line SOFLIC's can be established, since they utilise the compact support property of B-spline based fuzzy sets to ensure only a small number of fuzzy model parameters are updated at each iteration.

5.5 APPLICATION OF SOFLIC'S AND B-SPLINE NEUROCONTROLLERS TO AUTONOMOUS LAND VEHICLE CONTROL

The lateral dynamics of a conventionally steered road vehicle can be represented by the equations

$$m v \dot{\beta} = (m v^2 + \mu c_f l_f - \mu c_r l_r)\frac{\dot{\Psi}}{v} - (U_f + U_r + \mu c_f + \mu c_r)\beta$$

$$- (U_f + \mu c_f)\delta$$

$$I_z \ddot{\psi} = (c_f l_f - c_r l_r)\mu\beta - (c_f l_f^2 + c_r l_f^2)\frac{\mu\dot{\psi}}{v} + (U_f l_f + \mu c_f l_f)\delta \tag{5.5.1}$$

where for a 3500 kg. testbed vehicle, the parameters are:-

---

m = mass = 3500 kg; l_f = front wheels to C.G = 2.23m,

I_z = MI. = 12500 kg.m^{-1}; l_r = rear wheels to C.G = 1.02m

c_r = rear tyre side force coefficient = 120000 N/rad

c_f = front tyre side force coefficient = 50000 N/rad

U_f, u_r = front and rear circumferential forces (ignore)

β = vehicle side slip angle; ψ = vehicle heading, δ = steer angle.

The steering actuator is a first order lag with time constant of 0.35s; and actuator limits are $|\delta| \leq 40°$, $|\delta| \leq 6°\,s^{-1}$.

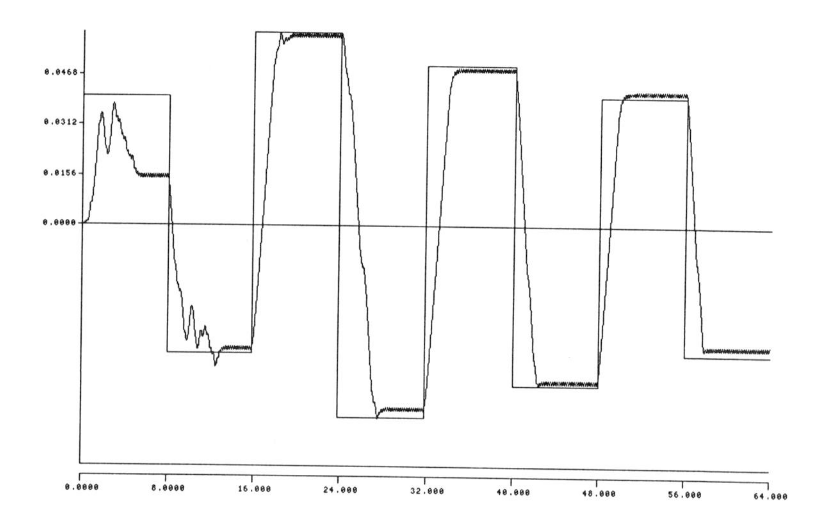

Figure 5.5 Lateral control at v = 5 m.s^{-1} (no a priori knowledge)

(i) Indirect SOFLIC lateral control

Eleven nonlinear discretisation levels were defined over the control and state variables (δ, x $\underline{\Delta}$ ψ/v, x, x), a forgetting factor of D = 0.9 with model update threshold of 0.4 were chosen (see Moore 1991 for details). Desired performance was specified in terms of a linear second order system with a natural frequenty of 5 rad.s^{-1} and damping ratio of 0.7. A controller sample rate of 10z was selected. For a fixed speed of 5m.s^{-1}, the controller was subject to a step demand, the maximum path curvature demanded represented a radius of curvature of 16m and lateral acceleration of 0.16g. the noise free transient reponse is shown in figure 5.5 for variable curvature demand. The intial poor response is due to the initially empty fuzzy rule base, some local relearning is required at each *different* curvature demand owing to the nonlinear dynamics of the vehicle. The vehicles velocity was then increased to motorway conditions of v = 20m.s^{-1}, demanded curvature and lateral accelerations at these conditions were respectively 125m and 0.32g. The SOFLIC controller response is shown in figure 5.6 based upon the *a priori* knowledge of driving at 5m.s^{-1}, again some local adaptation has occured.

Changing adhesion parameters, μ, adding sensor noise and wide distrubances produces similar responses, indicating substantial robustness in the SOFLIC algorithm (Moore, 1991).

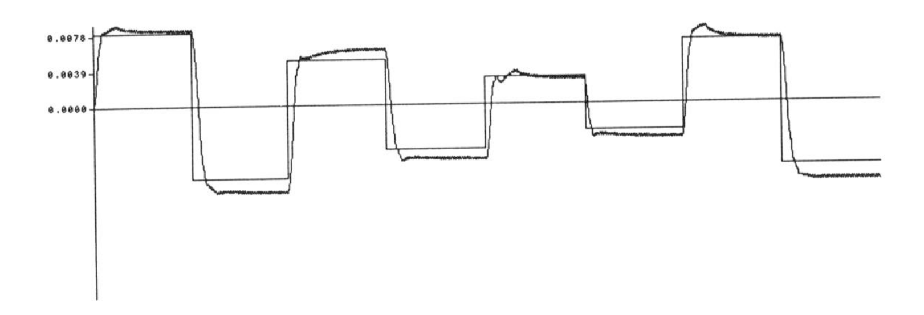

Figure 5.6 Lateral control at v = 20 m.s-1 (with 5m.s-1 model)

(ii) B-spline neurocontroller

For the same vehicle under the same dynamic conditions a B-spline neurocontroller consisting of the current curvature, current rate of change of curvature and the desired rate of change of curvature for the next iteration. The networks output is the desired steering angle δ. To initialise the network, all weights are zero and a training signal is provided by propagating the curvature error through an appropriate scaled forward plant model (represented as a Jacobian) in the form of a simple gain. This gain is selected to provide an approximate plant 'model' throughout the vehicle's operating envelope. Each input variable to the neurocontroller was covered by B-splines of order 2 (triangular sets), with a nonlinear knot strategy being employed. Fast initial convergence and long term stability was achieved by using a stochastic approximation NLMS weight learning algorithm with weight up date factor $\delta = 1/r$. the neurocontroller response for vehicle lateral control is shown in figure 5.7a, for a vehicle driving at a constant speed of $5 m.s^{-1}$, subject to random noise at the 10% level. Fast initial learning convergence is achieved, as well as an ability to generalise previously acquired knowledge. After only 20 repetitive learning cycles, the controller has learn't to fine tune control the vehicle as well as to filter out the random additive noise - see figure 5.7b.

5.6 DISCUSSION

Single layer associative memory neural nets such as Radial Basis Functions, CMAC, and B-spline networks have many desirable properties for neurocontrol : the learning algorithms are implementable in realtime, fast initial convergence and long term global convergence can be established, and local generalisation ensures that the learning proceeds stably. It has been shown that these algorithm s provide, under certain technical conditions, best approximations with arbitrarily small error to general continuous nonlinear functions. Whilst these networks (including SOFLIC's) are overall nonlinear, the output is linear in the set of adjustable parameters. Such an approach demands of the user some prior knowledge to optimally select the dimension (order) of the B-spline and knot placement strategy, or the choice of the RBF polynominal $\psi(\,.\,)$ and their centroids $\underline{c}$, however this knowledge is usually available in practical situations from the plant's physics, or operator experience, or from simulation studies. This representation also provides a natural framework for comparing neural nets with self-organising fuzzy logic systems and for generating formal rule update laws (or equivalently set confidences/membership functions) with proven stability and convergence rate conditions. However it should be noted that single layer networks increase the dimensionality of the input space (i.e. generate an increase in the network

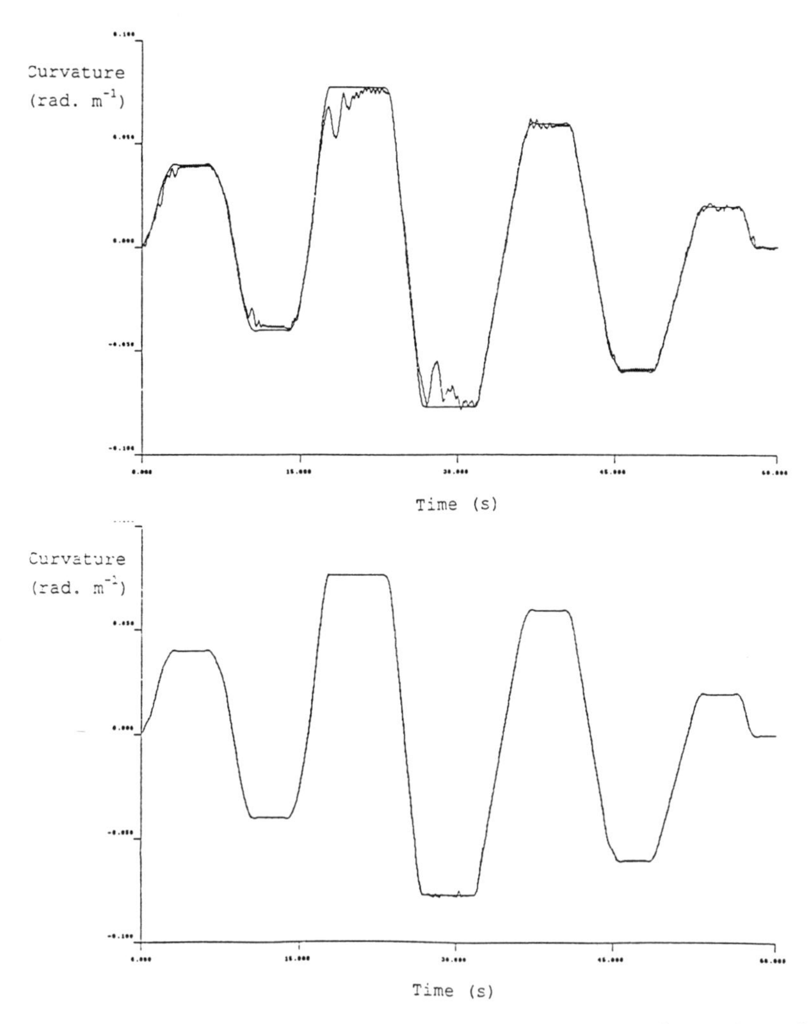

Figure 5.7 BSNN lateral control a. Initial learning, b. After 20 iterations

nodes or neurons) and associated increase in memory requirements - these algorithms are exponential in the dimension n of the input space.

It is possible to use static B-spline networks to implement guidance tasks such as path planning, corridor following, car parking, or in process modelling from sets of input/output data (Brown, 1990), where the decision surface is well knwon. B-splines have been effectively used for on-line dynamically constrained trajectory planning to produce safe and feasible vehicle velocity/position trajectories in complex obstacle fields (Harris et al, 1992). Regretfully space does not permit coverage here.

5.7 REFERENCES

1. Brown, M. (1990). Surface fitting algorithms for intelligent control. EEC Esprit II Panorama. Dept Soton TU.5.

2. Brown, M; and Harris, C.J. (1992). On line nonlinear time series prediction, EEC Esprit II Panorama Dept SotonTu 13.

3. Brown, M; and Harris, C.J. (1991). Fuzzy logic, neural networks and B-splines for intelligent control. Int. J. Math. Control and Information Theory. (OUP) (to appear).

4. Chen, S and Billings, S. (1992). Neural networks for nonlinear dynamic system modelling and identification. Int. J. Control - special issue on Intelligent Control (to appear).

5. Cotter, M. E. (1990); The Stone-Weierstrass theorem and its applications to neural nets. IEEE. Trans. Neural Networks Vol. I, pp290-295.

6. Cox, M.G.; Harris, P.M. and Jones, H.M. (1990). A knot placement strategy for least squares spline fitting based on the use of local polynomial approximations. In Algorithms for Approximation. Vol. II. Eds. J.C. Mason & M.G. Cox. Chapman & Hall (London). pp.37-45.

7. Harris, C.J.; Brown, M. and Moore, C.G. (1992). Intelligent Control: aspects of fuzzy logic and neural nets. World Scientific Press (Singapore).

8. Mamdani, E.H. (1974). Applications of fuzzy algorithms for control of simple dynamic plant. Proc. IEE. Vol. 121, No.12.

9. Mamdani, EH, and Assilian, S. (1975); An experiment in linguistic synthesis with a fuzzy logic controller . Int. J. Man. Mach. Stad. Vol. 7, pp.1-13.

10. Moore, C.G., and Harris, C.J. (1992). Indirect adaptive fuzzy control. Int. J. Control - special issue on Intelligent Control (to appear).

11. Moore, C.G. (1991). Indirect adaptive fuzzy control. Ph.D Thesis. University of Southampton.

12. Narendra, K.S.; and Parathasathy, K. (1990). Identification and control of dynamical systems using neural networks. IEEE. Trans. Neural Networks. Vol.I. No. 1. pp.4-27.

13. Pao, Y.H. (1989). Adaptive pattern recognition and neural networks. Addison Wesley Cp. (NY).

14. Procyk, T.J., and Mamdani, E.H. (1979). A linguistic self-organising process controller. Automatica, Vol.15, pp.15-30.

15. Sutton, R. and Jess, I.M. (1991). A design study of a self-organising fuzzy autopilot for ship control. Proc. Inst. Mech. Engrs. Vol.205. pp.35-47.

16. Zadeh, L.A. (1965). Fuzzy sets. J. Information and Control. Vol.8, pp.338-353. (also see same journal Vol.12 pp.94-102, (1968).

Chapter 6

Studies in neural network based control

K. J. Hunt and D. Sbarbaro

6.1 Introduction

Artificial neural networks can be used as a representation framework for modelling nonlinear dynamical systems. It is also possible to incorporate these nonlinear models within nonlinear feedback control structures. Several possibilities for modelling and control of nonlinear dynamical systems are studied in this paper. We present case studies illustrating the application of these techniques.

6.2 Representation and Identification

In this section we review the possibilities for using neural networks directly in nonlinear control strategies. In this context neural networks are viewed as a process modelling formalism, or even a knowledge representation framework; our knowledge about the plant dynamics and mapping characteristics is implicitly stored within the network.

The ability of networks to approximate nonlinear mappings is thus of central importance in this task. For this reason we first review a body of theoretical work which has characterised, from an approximation theory viewpoint, the possibilities of nonlinear functional approximation using neural networks.

Second, we discuss learning structures for training networks to represent forward and inverse dynamics of nonlinear systems. Finally, a number of control structures in which such models play a central role are reviewed. These established control structures provide a basis for nonlinear control using neural networks.

In the same way that transfer functions provide a generic representation for linear black-box models, ANNs potentially provide a generic representation for nonlinear black-box models.

6.2.1 Networks, Approximation and Modelling

The nonlinear functional mapping properties of neural networks are central to their use in control. Training a neural network using input-output data from a nonlinear plant can be considered as a nonlinear functional approximation problem. Approximation theory is a classical field of mathematics; from the famous Weierstrass Theorem [1] [2] it is known that polynomials, and many other approximation schemes, can approximate arbitrarily well a continuous function. Recently, considerable effort has gone into the application of a similar mathematical machinery in the investigation of the approximation capabilities of networks.

A number of results have been published showing that a feedforward network of the multilayer perceptron type can approximate arbitrarily well a continuous function [3, 4, 5, 6, 7]. To be specific, these papers prove that a continuous function can be arbitrarily well approximated by a feedforward network with only a single internal hidden layer, where each unit in the hidden layer has a continuous sigmoidal nonlinearity. The use of nonlinearities other than sigmoids is also discussed.

These results provide no special motivation for the use of networks in preference to, say, polynomial methods for approximation; both approaches share the 'Weierstrass Property'. Such comparitive judgements must be made on the basis of issues such as parsimony. Namely, do networks or polynomials require fewer parameters? For network approximators, key questions are how many layers of hidden units should be used, and how many units are required in each layer? Of course, the implementation characteristics of approximation schemes also provide a further basis for comparison (for example, the parallel, distributed, nature of networks is important). For the moment, however, we concentrate purely on approximation properties.

Although the results referred to above at first sight appear attractive they do not provide much insight into these practical questions. In fact, the degree of arbitrariness of the approximation achieved using a sigmoidal network with one hidden layer is reflected in a corresponding arbitrariness in the number of units required in the hidden layer; the results were achieved by placing no restriction on the number of units used. Cybenko himself says [4] "we suspect quite strongly that the overwhelming majority of approximation problems will require astronomical numbers of terms".

What is needed now is an indication of the numbers of layers/units required to achieve a specific degree of accuracy for the function being approximated. Some work along these lines is given in Chester [8]. That paper gives theoretical support to the empirical observation that networks with two hidden layers appear to provide higher accuracy and better generalisation than a single hidden layer network, and at a lower cost (i.e. fewer total processing units). Guidelines derived from a mix of theoretical and heuristic considerations are given in the introductory paper by Lippmann [9].

From the theoretical point of view the work of Kolmogorov [10] (see also [11]) did appear to throw some light on the problem of exact approximation [12, 13, 9]. Kolmogorov's theorem (a negative resolution of Hilbert's thirteenth problem) states that any continuous function of N variables can be computed using only linear summations and nonlinear but continuously increasing functions of only one variable.

In the network context the theorem can be interpreted as explicitly stating that to approximate any continuous function of N variables requires a network having $N(2N+1)$ units in a first hidden layer and $(2N+1)$ units in a second hidden layer. However, it has recently been pointed out [6, 14] that the practical value of this result is tenuous for a number of reasons:

1. Kolmogorov's Theorem requires a *different* nonlinear processing functions for each unit in the network.

2. The functions in the first hidden layer are required to be highly non-smooth. In practise this would lead to problems with generalisation and noise robustness.

3. The functions in the second hidden layer depend upon the function being approximated.

It is clear that these conditions are violated in the practical situations of interest here.

From the foregoing discussion it is clear that the property of approximating functions arbitrarily well is not sufficient for characterising good approximation schemes (since many schemes have this property), nor does this property help in justifying the use of one particular approximation scheme in preference to another. This observation has been deeply considered by Girosi and Poggio [15] (see also the expository paper [16]). These authors propose that the key property is not that of arbitrary approximation, but the property of *best approximation.* An approximation scheme is said to have this property if in the set of approximating functions there is one which has the minimum distance from the given function (a precise mathematical formulation is given in the paper). The first main result of their paper [15] is that multilayer perceptron networks *do not* have the best approximation property. Secondly, they prove that Radial Basis Function networks (for example, Gaussian networks) *do have* the best approximation property. Thus, although one must bear in mind the precise mathematical formulation of 'best', there is theoretical support for favouring RBF networks. Moreover, these networks may always be structured with only a single hidden layer and trained using linear optimisation techniques with a guaranteed global solution [17].

As with sigmoidal feedforward networks, however, there remain open questions regarding the network complexity required (i.e. the number of units). For RBF networks this question is directly related to the size of the training data. A related issue is that of choosing the centres of the basis functions (for results on automatically selecting the centres see Moody and Darken [18] and Chen *et el* [19]). Girosi and Poggio [15] and Broomhead and Lowe [17] discuss methods for achieving *almost best approximation* using restricted complexity RBF networks. This is discussed further in Sbarbaro and Gawthrop [20].

Thus, although RBF networks have the best approximation property, it is necessary to consider the cost of utlilising this approach. For high dimensional input spaces the number of nodes needed grows with the dimension of the input space [21, 22]. This is in contrast to sigmoids which slice up the space into feature regions and hence are more economical with the number of nodes.

In summary, a large body of theoretical results relating to approximation using neural networks exists. These results provide theoretically important possibility theorems and deep insight into the ultimate performance of networks. They are not constructive results which define the type and structure of a suitable network for a given problem. Restraint must therefore be exercised when citing these results in support of the practical application of neural networks.

6.2.2 Identification

Although a number of key theoretical problems remain, the results discussed above do demonstrate that neural networks have great promise in the modelling of nonlinear systems. Without reference to any particular network structure we now discuss architectures for training networks to represent nonlinear dynamical systems and their inverses.

An important question in system identification is that of system identifiability (see [23, 24, 25]) i.e. given a particular model structure, can the system under study be adequately represented within that structure? In the absence of such concrete theoretical results for neural networks we proceed under the assumption that all systems we are likely to study belong to the class of systems that the chosen network is able to represent.

Forward modelling

The procedure of training a neural network to represent the forward dynamics of a system will be referred to as *forward modelling*. A structure for achieving this is shown schematically in Figure 6.1. The neural network model is placed in parallel with the system and the error between the system and network outputs (the prediction error) is used as the network training signal. As pointed out by Jordan and Rumelhart [26] this learning structure is a classical supervised learning problem where the teacher (i.e. the system) provides target values (i.e. its outputs) directly in the output coordinate system of the learner (i.e. the network model). In the particular case of a multilayer perceptron type network straightforward backpropagation of the prediction error through the network would provide a possible training algorithm.

An issue in the context of control is the *dynamic* nature of the systems under study. One possibility is to introduce dynamics into the network itself. This can be done either using recurrent networks [27] or by introducing dynamic behaviour into the neurons (see [28]). A straightforward approach, and the one which for purposes of exposition will be followed here, is to augment the network input with signals corresponding to past inputs and outputs.

We assume that the system is governed by the following nonlinear discrete-time difference equation:

$$y^p(t+1) = f(y^p(t), \ldots, y^p(t-n+1); u(t), \ldots, u(t-m+1)) \qquad (6.1)$$

Thus, the system output y^p at time $t+1$ depends (in the sense defined by the nonlinear map f) on the past n output values and on the past m values of the input u. We concentrate here on the dynamical part of the system response; the model does not explicitly represent plant disturbances (for a method of including disturbances see Chen *et al* [19]). Special cases of the model (6.1) have been considered by

Narendra and Parthasarathy [29] (see also [30]). These authors consider particular cases where the system output is linear in either the past values of y^p or u.

An obvious approach for system modelling is to choose the input-output structure of the neural network to be the same as that of the system. Denoting the output of the network as y^m we then have

$$y^m(t+1) = \hat{f}(y^p(t), \ldots, y^p(t-n+1); u(t), \ldots, u(t-m+1)) \tag{6.2}$$

Here, $\hat{f}$ represents the nonlinear input-output map of the network (i.e. the approximation of f). Notice that the input to the network includes the past values of the *real* system output (the network has no feedback). This dependence on the system output is not included in the schematic of Figure 6.1 for simplicity. If we assume that after a suitable training period the network gives a good representation of the plant (i.e. $y^m \approx y^p$) then for subsequent post-training purposes the network output itself (and its delayed values) can be fed-back and used as part of the network input. In this way the network can be used independently of the plant. Such a network model is described by

$$y^m(t+1) = \hat{f}(y^m(t), \ldots, y^m(t-n+1); u(t), \ldots, u(t-m+1)) \tag{6.3}$$

The structure in (6.3) may also be used for training the network. This possibility has been discussed by Narendra [29, 30].

In the context of the identification of *linear* time invariant systems the two possibilities have been extensively considered by Narendra and Annaswamy [31]. The two structures have also been discussed in the signal processing literature (see Widrow and Stearns [32]). The structure of Equation (6.2) (referred to as the series-parallel model by Narendra) is supported in the identification context by stability results. On the other hand, (6.3) (referred to by Narendra as the parallel model) may be preferred when dealing with noisy systems since it avoids problems of bias caused by noise on the real system output [32, 33].

Inverse modelling

Inverse models of dynamical systems play a crucial role in a range of control structures. This will become apparent in Section 6.5. However, obtaining inverse models raises several important issues which will be discussed.

Conceptually the simplest approach is *direct inverse modelling* as shown schematically in part (a) of Figure 6.2 (this structure has also been referred to as *generalised inverse learning* [34]). Here, a synthetic training signal is introduced to the system. The system output is then used as input to the network. The network output is compared with the training signal (the system input) and this error is used to train the network. This structure will clearly tend to force the network to represent the inverse of the plant. However, there are drawbacks to this approach:

- The learning procedure is not 'goal directed' [26]; the training signal must be chosen to sample over a wide range of system inputs, and the actual operational inputs may be hard to define *a priori*. The actual goal in the control context is to make the system *output* behave in a desired way, and thus the training signal in direct inverse modelling does not correspond to the explicit goal.

- Second, if the nonlinear system mapping is not one-one then an incorrect inverse can be obtained.

The first point above is strongly related with the general concept of ***persistent excitation***; the importance of the inputs used to train learning systems is widely appreciated. In the adaptive control literature conditions for ensuring persistent excitation which will result in parameter convergence are well established (see, for example, Åström and Wittenmark [35] and the references therein). For neural networks, methods of characterising persistent excitation are highly desirable. A preliminary discussion on this question can be found in Narendra [30].

A second approach to inverse modelling which aims to overcome these problems is known as *specialised inverse learning* [34] (somewhat confusingly, Jordan and Rumelhart [26] refer to this structure as *forward modelling*). The specialised inverse learning structure is shown in part (b) of Figure 6.2. In this approach the network inverse model precedes the system and receives as input a training signal which spans the desired operational output space of the controlled system (i.e. it corresponds to the system reference or command signal). This learning structure also contains a trained forward model of the system (for example, a network trained as described in Section 6.2.2) placed in parallel with the plant. The error signal for the training algorithm in this case is the difference between the training signal and the system output (it may also be the difference between the training signal and the forward model output in the case of noisy systems; this obviates the need for the real system in the training procedure which is important in situations where using the real system is not viable). Jordan and Rumelhart [26] show that using the real system output can produce an exact inverse even when the forward model is inexact; this is not the case when the forward model output is used. The error may then be propagated back through the forward model and then the inverse model; only the inverse network model weights are adjusted during this procedure. Thus, the procedure is effectively directed at learning an identity mapping across the inverse model and the forward model; the inverse model is learned as a side effect [26].

In comparison with direct inverse modelling the specialised inverse learning approach has the following features:

- The procedure is goal directed since it is based on the error between desired system outputs and actual outputs. In other words, the system receives inputs during training which correspond to the actual operational inputs it will subsequently receive.

- In cases in which the system forward mapping is not one-one a particular inverse will be found (Jordan and Rumelhart [26] discuss ways in which learning can be biased to find particular inverse models with desired properties).

We now consider the input-output structure of the network modelling the system inverse. From Equation (6.1) the inverse function f^{-1} leading to the generation of $u(t)$ would require knowledge of the future value $y^p(t+1)$. To overcome this problem we replace this future value with the value $r(t+1)$ which we assume is available at time t. This is a reasonable assumption since r is typically related to the reference signal which is normally known one step ahead. Thus, the nonlinear input-output relation of the network modelling the plant inverse is

$$u(t) = \widehat{f^{-1}}(y^p(t), \ldots, y^p(t-n+1); r(t+1); u(t-1), \ldots, u(t-m+1)) \qquad (6.4)$$

i.e. the inverse model network receives as inputs the current and past system outputs, the training (reference) signal, and the past values of the system input. In cases where it is desirable to train the inverse without the real system (as discussed above) the values of y^p in the above relation are simply replaced by the forward model outputs y^m.

6.3 Gaussian Networks

The networks considered here are Gaussian feedforward networks with one hidden layer. We consider for the moment a network having many inputs and a single output (by duplication of this structure the approach can be generalised easily to multi-output systems). The number of input units, d, corresponds to the dimension of the network input vector, $u_{network} \in \Re^d$.

The *linear* output unit is fully connected to the hidden units; the network output (the activation value of the output unit) is a weighted sum of the activation levels of the N hidden units:

$$y_{network} = \sum_{h=1}^{N} w_{yh} o_h \qquad (6.5)$$

Here, w_{yh} is the connection weight between hidden unit h and the network output, $y_{network}$. o_h is the output (activation level) of hidden unit h.

In Gaussian units the activation level of a hidden unit depends only on the distance between the input vector $u_{network}$ and the centre of the Gaussian function of that unit. The centre of the function for hidden unit h is denoted by the vector $\mu_h \in \Re^d$. For hidden unit h,

$$o_h = \exp\left(\frac{||u_{network} - \mu_h||^2}{-2\sigma_h^2}\right) \qquad (6.6)$$

where σ_h is the width of the Gaussian function of unit h.

For nonlinear dynamic systems modelling purposes we assume that the plant is governed by the following nonlinear difference equation:

$$y^p(t+1) = f(y^p(t), \ldots, y^p(t-n+1); u(t), \ldots, u(t-m+1)) \qquad (6.7)$$

where $y^p(.)$ is the plant output and $u(.)$ the input.

Following Equation (6.2) we select the structure of the neural network to be trained to represent the plant as

$$y^m(t+1) = \hat{f}(y^p(t), \ldots, y^p(t-n+1); u(t), \ldots, u(t-m+1)) \qquad (6.8)$$

Notice here that the input to the network includes the past values of the *real* plant output (the network has no feedback).

The nonlinear relation of the connectionist network modelling the plant inverse is given by

$$u(t) = \widehat{f^{-1}}(y^m(t), \ldots, y^m(t-n+1); r(t+1); u(t-1), \ldots, u(t-m+1)) \quad (6.9)$$

Here, we base the inverse model on the plant model (as indicated by the subscript m) rather than on the true plant (c.f Equation (6.4)).

The inversion problem can be formulated as the problem of solving an operator equation; given the operator N represented by a neural network mapping $[y(t), \ldots, y(t-n), u^*(t), \ldots, u(t-m)]$ to $y(t+1) = y^*$;

$$Nu^* = y^*$$

Then, given $[y(t), \ldots, y(t-n), u(t-1), \ldots, u(t-m)]$ compute $u^*(t)$ to y^*. In [36] two basic methods for solving the operator equation for u^* are described. One uses the contraction principle, the second involves the use of sensitivity functions and Newton's iteration method.

The advantages of the connectionist approach are firstly it provides differentiable models to be used as sensitivity functions and, secondly, association capabilities. In this way the information about u^* is stored in the network and it is not necessary to start the iterations from the same point every time. This network will finally represent the inverse of the model.

6.4 Learning Algorithms

In Section 6.2.2 we outlined the two architectures used for plant modelling and plant inverse modelling. These are necessary in the IMC formulation. In this section we present the learning laws used for training the networks.

6.4.1 Modelling the plant

The plant is modelled using a network described by

$$y^m(t+1) = \sum_{i=1}^{N^m} c_i^m K_i^m \quad (6.10)$$

where

$$K_i^m = e^{-d_i^m(x^m(t), x_i^m, \Delta^m)} \quad (6.11)$$

Here, the m superscript indicates a variable related with the plant model. We choose the structure of the plant model to be the same as that of the plant i.e. the model output is a nonlinear function of the present and past plant outputs, and the present and past plant inputs. The model input vector $x^m(t)$ is thus given as

$$x^m(t) = [y^p(t), \ldots, y^p(t-n+1), u(t), \ldots, u(t-m+1)]^T$$

We denote the centre of the Gaussian function of hidden unit i as

$$x_i^m = [y_{i,1}, \ldots, y_{i,n}, u_{i,1}, \ldots, u_{i,m}]^T$$

The parameters x_i^m and Δ^m are fixed to meet the interpolation conditions, i.e. the x_i^m are distributed uniformly over the input space and Δ^m is adjusted such that $\sum_{i=1}^{N^m} K_i = constant$ over the input space. There are other possibilities for this [37].

The parameter vector c_i^m is adjusted to minimize the mean square error between the real plant and the model. That is,

$$c_i^m(t+1) = c_i^m(t) + \alpha K_i^m(y^p(t+1) - y^m(t+1)) \tag{6.12}$$

Here, α is a gain parameter. Using standard linear systems theory it can be shown that if the plant can be modelled as (6.10), the least mean square solution can be found by (6.12) [20].

6.4.2 Inverse model identification

If the model of the plant is invertible then the inverse of the plant can be approximated in a similar way to the plant. This model is then used as the controller. For reasons described in Section 6.4.2 we choose to use the plant *model* inverse rather than the inverse of the real plant. We utilise a second network described by

$$u(t) = \sum_{i=1}^{N^C} c_i^C K_i^C \tag{6.13}$$

where,

$$K_i^C = e^{-d_i^C(x^C(t), x_i^C, \Delta^C)} \tag{6.14}$$

Here, the C superscript indicates a variable related with the controller. The inverse of the function f in Equation (6.7) (required to obtain $u(t)$) depends upon the future plant output value $y^p(t+1)$. In order to obtain a realisable approximation we replace this value by the controller input value r. Finally, since we actually require to approximate the inverse of the plant model (as opposed to the plant itself), we define the controller network input vector $x^C(t)$ as

$$x^C(t) = [y^m(t), \ldots, y^m(t-n+1), r(t+1), u(t-1) \ldots, u(t-m+1)]^T$$

Here, the future value $r(t+1)$ is obtained at time t by suitable definition of the IMC filter F. The centre of the Gaussian function of hidden unit i is given by

$$x_i^C = [y_{i,1}, \ldots, y_{i,n}, r_i, u_{i,2} \ldots, u_{i,m}]^T$$

Non-iterative methods

The architecture used to adjust c_i^C is similar to the specialised learning architecture presented by Psaltis *et al* [34] (the difference being that here we use the plant model, rather than the plant itself). The parameters in c_i^C are adjusted to minimize the mean square error between the output of the model and the input of the controller. This leads to the following learning algorithm:

$$c_i^C(t+1) = c_i^C(t) + \alpha K_i^C(r(t+1) - y^m(t+1))\frac{\partial y^m(t+1)}{\partial u(t)} \tag{6.15}$$

Here, if the real plant were used in the learning procedure (as in [34]) then $\frac{\partial y(t+1)}{\partial u(t)}$ would require to be estimated. This can be done using first order differences [34] changing each input to the plant slightly at the operating point and measuring the change at the output. By using the plant model, however, the derivatives can be calculated explicitly. From Equation (6.10) we obtain

$$\frac{\partial y^m(t+1)}{\partial u(t)} = -2\Delta^m \sum_{i=1}^{N^m} c_i^m K_i^m(u(t) - u_{i,1}) \tag{6.16}$$

Proposition 1 *The learning algorithm defined by Equation (6.15) converges to a global minimum of the index defined by*

$$J = \sum_j (r(j+1) - y^m(j+1))^2 \tag{6.17}$$

if the system is monotonically increasing with respect to $u(t)$.

Proof:

See Hunt and Sbarbaro [38]. ◇◇

Another approach involves the use of a synthetic signal [33]. This leads to the so called general learning architecture [34] as shown in Figure 6.2 (direct method). In this case the adaptation law for the weights does not depend on the derivatives of the plant:

$$c_i^C(t+1) = c_i^C(t) + \alpha K_i^C(S_s - u(t)) \tag{6.18}$$

Here, S_s is the synthetic signal.

Proposition 2 *If the system is invertible the algorithm defined by Equation (6.18) converges to the best approximation of the inverse in the least square sense.*

Proof:

If the system is invertible then there exists an injective mapping which represents the inverse. Thus, from linear systems theory the algorithm defined by (6.18) converges to the least squares error [20]. ◇◇

As pointed out in Psaltis *et al* [34] the specialised method allows the training of the inverse network in a region in the expected operational range of the plant. On the other hand, the generalised training procedure produces an inverse over the whole operating space. Psaltis *et al* suggest a hybrid training procedure where the specialised and generalised architectures are combined.

Iterative methods

Iterative methods make use of a plant model to calculate the inverse. In this case a recursive method is used to find the inverse of the model in each operating point. This method is useful in singular systems which satisfy the invertibility conditions outlined earlier only locally and not for the whole operating space. This approach can also be used when it is necessary to have small networks due to memory or processing limitations. In this case the restricted accuracy of the trained network can be enhanced by using the network to provide stored initial values for the iterative method, establishing a compromise between speed of convergence and storing capacities.

At time t, the objective is to find an input u which will produce a model output $y^m(t+1)$ equal to $r(t+1)$. It is possible to use the method of successive substitution:

$$u^{n+1}(t) = u^n(t) + \gamma(r(t+1) - y^m(t+1))$$

where γ is a weight to be chosen.

According to the small gain theorem [39], the inverse operator is input-output stable if the product of the operator gains in the loop is less than 1:

$$\|I\|\|I - \gamma f\| \leq 1$$

The initial value $u^0(t)$ can be stored in a connectionist network.

6.5 Control Structures

Models of dynamical systems and their inverses have immediate utility for control. In the control literature a number of well established and deeply analysed structures for the control of nonlinear systems exist; we focus on those structures having a direct reliance on system forward and inverse models. We assume that such models are available in the form of neural networks which have been trained using the techniques outlined above.

In the literature on neural network architectures for control a large number of control structures have been proposed and used; it is beyond the scope of this work to provide a full survey of all architectures used. In the sequel we give particular emphasis to those structures which, from the mainstream control theory viewpoint, are well established and whose properties have been deeply analysed. First, we briefly discuss two direct approaches to control: supervised control and direct inverse control.

6.5.1 Supervised control

There are many control situations where a human provides the feedback control actions for a particular task and where it has proven difficult to design an automatic controller using standard control techniques (e.g. it may be impossible to obtain an analytical model of the controlled system). In some situations it may be desirable to design an automatic controller which mimics the action of of the human (this has been called *supervised control* [40]).

A neural network provides one possibility for this (as an alternative approach expert systems can be used to provide the knowledge representation and control formalisms). Training the network is similar in principle to learning a system forward model as described above. In this case, however, the network input corresponds to the sensory input information received by the human. The network target outputs used for training correspond to the human control input to the system. This approach has been used in the standard pole-cart control problem [41], among others.

6.5.2 Direct inverse control

Direct inverse control utilises an inverse system model. The inverse model is simply cascaded with the controlled system in order that the composed system results in an identity mapping between desired response (i.e. the network inputs) and the controlled system output. Thus, the network acts directly as the controller in such a configuration. Direct inverse control is common in robotics applications; the compilation book [42] provides a number of examples.

Clearly, this approach relies heavily on the fidelity of the inverse model used as the controller. For general purpose use serious questions arise regarding the robustness of direct inverse control. This lack of robustness can be attributed primarily to the absence of feedback. This problem can be overcome to some extent by using on-line learning: the parameters of the inverse model can be adjusted on-line.

6.5.3 Model reference control

Here, the desired performance of the closed-loop system is specified through a stable reference model M, which is defined by its input-output pair $\{r(t), y^r(t)\}$. The control system attempts to make the plant output $y^p(t)$ match the reference model output asymptotically, i.e.

$$\lim_{k\to\infty} \|y^r(t) - y^p(t)\| \leq \epsilon$$

for some specified constant $\epsilon \geq 0$. The model reference control structure for nonlinear systems utilising connectionist models is shown in Figure 6.3 [29]. In this structure the error defined above is used to train the network acting as the controller. Clearly, this approach is related to the training of inverse plant models as outlined above. In the case when the reference model is the identity mapping the two approaches coincide. In general, the training procedure will force the controller to be a 'detuned' inverse, in a sense defined by the reference model. Previous and similar work in this area has been done considering linear in the control structures [43].

6.5.4 Internal model control

In Internal Model Control (IMC) the role of system forward and inverse models is emphasised [44]. In this structure a system forward and inverse model are used directly as elements within the feedback loop. IMC has been thoroughly examined and shown to yield transparently to robustness and stability analysis [45]. Moreover, IMC extends readily to nonlinear systems control [36].

In internal model control a system model is placed in parallel with the real system. The difference between the system and model outputs is used for feedback purposes. This feedback signal is then processed by a controller subsystem in the forward path; the properties of IMC dictate that this part of the controller should be related to the system inverse (the nonlinear realisation of IMC is illustrated in Figure 6.4).

Given network models for the system forward and inverse dynamics the realisation of IMC using neural networks is straightforward [38]; the system model M and the controller C (the inverse model) are realised using the neural network models as shown in Figure 6.5. The subsystem F is usually a linear filter which can be designed to introduce desirable robustness and tracking response to the closed-loop system.

It should be noted that the *implementation* structure of IMC is limited to open-loop stable systems. However, the technique has been widely applied in process control. From the control theoretic viewpoint there is strong support for the IMC approach. Examples of the use of neural networks for nonlinear IMC can be found in Hunt and Sbarbaro [38].

The IMC structure is now well known and has been shown to underlie a number of control design techniques of apparently different origin [44]. IMC has been shown to have a number of desirable properties; a deep analysis is given by Morari and Zafiriou [45]. Here, we briefly summarise these properties.

The nonlinear IMC structure is shown in Figure 6.4 (in this section we follow Economou *et al* [36]). Here, the nonlinear operators denoted by P, M and C represent the plant, the plant model, and the controller, respectively. The operator F denotes a filter, to be discussed in the sequel. The double lines used in the block diagram emphasise that the operators are nonlinear and that the usual block diagram manipulations do not hold.

The important characteristics of IMC are summarised with the following properties:

(P.1) Assume that the plant and controller are input-output stable and that the model is a perfect representation of the plant. Then the closed-loop system is input-output stable.

(P.2) Assume that the inverse of the operator describing the plant model exists, that this inverse is used as the controller, and that the closed-loop system is input-output stable with this controller. Then the control will be perfect i.e. $y^p = y^s$.

(P.3) Assume that the inverse of the steady state model operator exists, that the steady state controller operator is equal to this, and that the closed-loop system is input-output stable with this controller. Then offset free control is attained for asymptotically constant inputs.

The IMC structure provides a direct method for the design of nonlinear feedback controllers. According to the above properties, if a good model of the plant is available, the closed-loop system will achieve exact setpoint following despite unmeasured disturbances acting on the plant.

Thus far, we have not described the role of the filter F in the system. The discussion so far has considered only the idealised case of a perfect model, leading to perfect control. In practice, however, a perfect model can never be obtained. In addition, the infinite gain required by perfect control would lead to sensitivity problems under model uncertainty. The filter F is introduced to alleviate these problems. By suitable design, the filter can be selected to reduce the gain of the feedback system, thereby moving away from the perfect controller. This introduces robustness into the IMC structure. A full treatment of robustness and filter design for IMC is given in Morari and Zafiriou [45].

The significance of IMC, in the context of this work, is that the stability and robustness properties of the structure can be analysed and manipulated in a transparent manner, even for nonlinear systems. Thus, IMC provides a general framework for nonlinear systems control. Such generality is not apparent in alternative approaches to nonlinear control.

A second role of the filter is to project the signal e into the appropriate input space for the controller.

We propose a two step procedure for using neural networks directly within the IMC structure. The first step involves training a network to represent the plant response. This network is then used as the plant model operator M in the control structure of Figure 6.5. The architecture shown in Figure 6.1 provides the method for training a network to represent the plant. Here, the error signal used to adjust the network weights is the difference between the plant output and the network output. Thus, the network is forced towards copying the plant dynamics. Full details of the learning law used here are given in Section 6.4.1.

Following standard IMC practice (guided by Property P.2 above) we select the controller as the plant inverse model. The second step in our procedure is therefore to train a second network to represent the inverse of the plant. To do this we use one of the architectures shown in Figure 6.2. Here, for reasons explained in Section 6.4.2 (where full details of the learning law are given), we employ the plant model (obtained in the first learning step) in the inverse learning architecture rather than the plant itself. For inverse modelling the error signal used to adjust the network is defined as the difference between the (inverse modelling) network input and the plant model output. This tends to force the transfer function between these two signals to unity i.e. the network being trained is forced to represent the inverse of the plant model. Having obtained the inverse model in this way this network is used as the controller block C in the control structure of Figure 6.5.

The final IMC architecture incorporating the trained networks is shown in Figure 6.5.

6.5.5 Predictive control

The receding horizon control approach can be summarised by the following steps:

1.- Predict the system output over the range of future times,

2.- Assume that the future desired outputs are known,

3.- Choose a set of future controls, $\hat{u}$, which minimize the future errors between the predicted future output and the future desired output,

4.- Use the first element of $\hat{u}$ as a current input and repeat the whole process at the next instant.

It has been shown that this technique can stabilize linear systems [46] and nonlinear systems as well [47].

The objective is to calculate the control such that the error over a future horizon is minimized, i.e. we consider a cost function of the following form:

$$J_{N_1,N_2,N_u} = \sum_{k=i+N_1}^{i+N_2} (y^r(k) - y^m(k))^2 + \sum_{k=i}^{i+N_u} \lambda_k(\Delta u(k))^2 \tag{6.19}$$

where y^r represents the output of the reference model (i.e. the desired output) and y^m the output of the plant model. The first term of the cost function is a measure of the distance between the model prediction and the desired future trajectory. The second term penalizes excessive movement of the manipulated variable.

The structure of the connectionist controller is shown in Figure 6.6. This is similar to a learning control structure previously proposed by Ersü and Tolle [48]. Those authors, however, consider only a one-step-ahead optimisation criterion. In our approach we propose the use of a trained connectionist model of the nonlinear plant M to predict the future plant outputs. This model is used to provide the prediction data to the optimisation algorithm. In principle, this optimised control value can then be applied to the plant. However, in a further step we also use this signal to train a controller network C, as shown in the figure. Training C in this way provides two important options: either to generate a useful starting value for the optimisation algorithm, or to act as a stand-alone feedback controller without recourse to the optimisation step. Of course, training C in the manner shown leads to the generation of an approximate plant inverse, depending on the cost function weights.

Statement of the problem

Consider a plant described as

$$\begin{aligned} y^p(t+1) &= f(y^p(t),\ldots,y^p(t-n+1); u(t),\ldots,u(t-m+1)) \quad t>i \\ [y^p(t),\ldots,y^p(t-n+1), u(t),\ldots,u(t-m+1)] &\in Z_t \subset R^{n+m} \quad t>i \\ y(i) &= a \end{aligned} \tag{6.20}$$

where f is a function defined as $f: R^{n+m} \to R$, such that the *moving horizon cost* at time i is

$$J_i = \sum_{k=i+N_1}^{i+T-1} \|e(k)\|^2 + \sum_{k=i}^{i+N_u} \|\Delta u(k)\|^2_{\lambda_i} \tag{6.21}$$

Here, $e(k)$ is defined as $e(k) = y^r(k) - y^m(k)$, $\Delta u(k) = u(k) - u(k-1)$, and the reference model is described by

$$\begin{aligned} y^r(t+1) &= g(y^r(t),\ldots,y^r(t-n+1); r(t),\ldots,r(t-m+1)) \\ [y^r(t),\ldots,y^r(t-n+1), r(t),\ldots,r(t-m+1)] &\in Z^r_t \subset R^{n+m} \\ t>i \quad y^r(i) &= b \end{aligned} \tag{6.22}$$

The problem is to determine a sequence $\{u(k)\}_{k=i}^{i+N_u}$ which minimizes the cost given by (6.21), subject to (6.20), (6.22) and the alternative constraints

$$i \leq k \leq i+T-1 \quad , \quad y^p(i+T) = y^r(i+T) \tag{6.23}$$

This means that the reference model must have a settling time less than or equal to T, that is $y^r(i+T) = y^r_{ss}$. The demonstration of existence of the solution, stability of the closed loop system, and robustness are found in [49], [47], and [50].

Minimizing the functional

To minimize the functional (6.21) a simple gradient algorithm was used, although more efficient, but at the same time more complex, algorithms can be applied. In this case the derivatives of the output against an input variable are estimated from the model and used to calculate the gradient in each iteration:

$$u|_{i+k}^{i+N_u} = u|_{i+k}^{i+N_u} + \eta(\nabla_{u|_{i+k}^{i+N_u}} I_y + \nabla_{u|_{i+k}^{i+N_u}} I_u) \tag{6.24}$$

where η fixes the step of the gradient, $I_y = \sum_{k=i+N_1}^{i+T-1}(y^r(k) - y^m(k))^2$, $I_u = \sum_{k=i}^{i+N_u} \lambda_k(\Delta u(k))^2$, and noting $N_2 = T-1$. The gradients are calculated as

$$\nabla_{u|_{i+k}^{i+N_u}} I_y = \begin{pmatrix} \frac{\partial y^m(i+N_1)}{\partial u(i)} & \cdots & \frac{\partial y^m(i+N_2)}{\partial u(i)} \\ \vdots & \ddots & \vdots \\ 0 & \cdots & \frac{\partial y^m(i+N_2)}{\partial u(i+N_u)} \end{pmatrix} \begin{pmatrix} y^r(i+N_1) - y^m(i+N_1) \\ \vdots \\ y^r(i+N_2) - y^m(i+N_2) \end{pmatrix} \tag{6.25}$$

$$\nabla_{u|_{i+k}^{i+N_u}} I_u = \begin{pmatrix} \lambda_1 & -\lambda_1 & \cdots & 0 \\ 0 & \lambda_2 & \cdots & 0 \\ \vdots & \vdots & \ddots & \vdots \\ 0 & 0 & \cdots & \lambda_{N_u} \end{pmatrix} \begin{pmatrix} \Delta u(i) \\ \Delta u(i+1) \\ \vdots \\ \Delta u(i+N_u) \end{pmatrix} \tag{6.26}$$

Representing the model as a connectionist network $y^m(t+1)$ is given by

$$y^m(t+1) = \sum_{i=1}^{N^m} c_i^m K_i^m \tag{6.27}$$

Here, N^m represents the number of units in a connectionist network, c_i linear coefficients, and K_i^m is a Gaussian function defined as

$$e^{-d_i^m(x^m(t), x_i^m, \Delta^m)}$$

where

$$d_i^m(x^m(t), x_i^m, \Delta_i) = (x^m(t) - x_i^m)^T \Delta_i (x^m(t) - x_i^m). \tag{6.28}$$

$x^m(t) = [y^m(t), \ldots, y^m(t-n+1), u(t), \ldots, u(t-m+1)]^T$ is the input vector for the network, and x_i^m represents the centre of the unit i. Δ_i^m is a diagonal matrix with σ in its diagonal.

To calculate the partial derivatives the followings formulas were used for each control action k with $k = 1 \ldots N_u$ and $j = N_1 \ldots (T-1) - 1$:

$$D_{k,j} = \frac{\partial y^m(i+j)}{\partial u(k)} =$$

$$\begin{cases} \sum_{l=1}^{N} c_l^m K^m(x^m(i+j), x_l^m, \Delta^m)(u^m(k) - u_l^m)\Delta^m & k = j \\ D_{k,j-1} \sum_{l=1}^{N} c_l^m K^m(x^m(i+j), x_l^m, \Delta^m)(y^m(i+j) - y_l^m)\Delta^m & k < j \\ 0 & k > j \end{cases} \tag{6.29}$$

At the end of the optimisation process it is necessary to verify that the condition $y^r(k+N_2) = y^m(k+N_2)$ is met.

For every starting point in the optimisation process, $x^m(t) = [y(t), \ldots, y(t-n+1), u(t), \ldots, u(t-m+1)]$, there is an associated optimal value of u_t. Thus, u_t can be expressed as

$$u_t = N_c(x_t, r_t) \tag{6.30}$$

where N_c is an operator to be represented by a connectionist network. This network becomes the controller C in our overall control structure.

Effect of the disturbances

It is important to include a reasonable representation of likely disturbances since the correct controller structure can then be deduced. As pointed out in [51], it should not be necessary to force the controller to have integral action, but rather this structure should arise naturally from reasonable assumptions about the dynamics of the controlled system. The general prediction model used to calculate the control signal can include an estimate of a step-like disturbance. The general model is given by

$$y^m(t+1) = \sum_{i=1}^{N^m} c_i^m K_i^m + \hat{d}(t)$$

where

$$\begin{aligned} \hat{d}(t+1) &= \hat{d}(t) & ,t > i \\ &= (y^p(t) - y^m(t)) & ,t = i \end{aligned} \tag{6.31}$$

6.6 Example - pH Control

pH control is an old problem involving significant nonlinearities and uncertainties. These features can be so severe that classical linear feedback control does not always achieve satisfactory performance. In its simplest form the process consists of a stirred tank in which waste water from a plant is neutralised using a reagent, usually a strong acid or a strong base (see Figure 6.7). The system is commonly modelled as the combination of a linear dynamic system followed by a nonlinear block representing the process titration curve. The nonlinear function representing the titration curve is shown in Figure 6.8.

In this case a neural network model was developed for the nonlinear system. The network was trained to learn the relation between the control variable and the process output. After presenting a set of training data consisting of 120 points (distributed over the operational range of the process variables) an absolute model mismatch error of 5% was obtained.

As discussed above, once the network has been trained to represent the nonlinear system it can be included in a nonlinear control loop. For this application we chose to utilise the Internal Model Control and receding horizon control structures; these approaches directly incorporate a model of the process in the feedback loop. The neural network was therefore directly incorporated as part of the nonlinear controllers.

The set-point response of the system for the controlled pH and the corresponding control variable are shown in the figures: the results for IMC are shown in Figure 6.9, the results for receding horizon control with $\lambda = 0$ in Figure 6.10, and the results for receding horizon control with $\lambda = 1$ in Figure 6.11.

It can be seen that the nonlinear neural network controllers provide very close tracking performance over the range of pH variation; this is a considerable improvement over a linear controller on this loop.

6.7 Acknowledgements

Kenneth Hunt currently holds a Royal Society of Edinburgh Personal Research Fellowship. Daniel Sbarbaro is supported by a Chilean government research scholarship.

Bibliography

[1] J. Burkill and H. Burkill, *A Second Course in Mathematical Analysis.* Cambridge, England: Cambridge University Press, 1970.

[2] W. Rudin, *Principles of Mathematical Analysis, 3rd Edition.* Auckland: McGraw-Hill, 1976.

[3] G. Cybenko, "Continuous valued neural networks with two hidden layers are sufficient," *Technical Report, Department of Computer Science, Tufts University*, 1988.

[4] G. Cybenko, "Approximation by superpositions of a sigmoidal function," *Math. Control Signal Systems*, vol. 2, pp. 303–314, 1989.

[5] K. I. Funahashi, "On the approximate realization of continuous mappings by neural networks," *Neural Networks*, vol. 2, pp. 183–192, 1989.

[6] K. Hornik, M. Stinchcombe, and H. White, "Multilayer feedforward networks are universal approximators," *Neural Networks*, vol. 2, pp. 359–366, 1989.

[7] S. M. Carrol and B. W. Dickinson, "Construction of neural nets using the Radon transform," in *Proc. IJCNN*, 1989.

[8] D. Chester, "Why two hidden layers are better than one," in *IEEE Int.Joint Conf. on Neural Networks, IJCNN'90*, pp. I 265–268, 1990.

[9] R. P. Lippmann, "An introduction to computing with neural nets," *IEEE ASSP Magazine*, vol. 4, pp. 4–22, 1987.

[10] A. N. Kolmogorov, "On the representation of continuous functions of several variables by superposition of continuous functions of one variable and addition," *Dokl. Akad. Nauk SSSR*, vol. 114, pp. 953–956, 1957.

[11] G. G. Lorentz, *Mathematical Developments Arising from Hilbert's Problems*, vol. 2, pp. 419–430. American Mathematical Society, 1976.

[12] T. Poggio, *Physical and Biological Processing of Images*, pp. 128–153. Springer-Verlag, 1982.

[13] R. Hecht-Nielsen, "Kolmogorov's mapping neural network existence theorem," in *Proc. IJCNN*, 1987.

[14] F. Girosi and T. Poggio, "Representation properties of networks: Kolmogorov's theorem is irrelevant," *Neural Computation*, vol. 1, pp. 465–469, 1989.

[15] F. Girosi and T. Poggio, "Networks and the best approximation property," *Biological Cybernetics*, vol. 63, pp. 169–176, 1990.

[16] T. Poggio and F. Girosi, "Networks for approximation and learning," *Proceedings of the IEEE*, vol. 78, pp. 1481–1497, 1990.

[17] D. S. Broomhead and D. Lowe, "Multivariable functional interpolation and adaptive networks," *Complex Systems*, vol. 2, pp. 321–355, 1988.

[18] J. Moody and C. Darken, "Fast-learning in networks of locally-tuned processing units," *Neural Computation*, vol. 1, pp. 281–294, 1989.

[19] S. Chen, S. A. Billings, C. F. Cowan, and P. M. Grant, "Practical identification of NARMAX models using radial basis functions," *Int. J. Control*, vol. 52, pp. 1327–1350, 1990.

[20] D. G. Sbarbaro and P. J. Gawthrop, "Self-organization and adaptation in gaussian networks," in *9th IFAC/IFORS symposium on identification and system parameter estimation. Budapest, Hungary*, pp. 454–459, 1991.

[21] E. Hartman and J. D. Keeler, "Predicting the future with semi-local units," *Neural Computation*. To appear.

[22] A. S. Weigand, B. A. Huberman, and D. E. Rumelhart, "Predicting the future: a connectionist approach," *Int. J. Neural Systems*, vol. 3, p. 193, 1990.

[23] L. Ljung and T. Söderström, *Theory and Practice of Recursive Identification*. London: MIT Press, 1983.

[24] L. Ljung, *System Identification: Theory for the User*. New York: Prentice-Hall, 1987.

[25] T. Söderström and P. Stoica, *System Identification*. Hemel Hempstead: Prentice-Hall, 1989.

[26] M. I. Jordan and D. E. Rumelhart, "Forward models: supervised learning with a distal teacher," *Submitted*, pp. 1–49, 1991.

[27] R. Żbikowski, "Dynamics identification and control with continuous neural networks," *Internal report, Control Group, University of Glasgow*, 1991. Submitted for publication.

[28] M. J. Willis, C. D. Massimo, G. A. Montague, M. T. Tham, and A. J. Morris, "On artificial neural networks in process engineering," *Proc. IEE Pt. D*, vol. 138, pp. 256–266, 1991.

[29] K. S. Narendra and K. Parthasarathy, "Identification and control of dynamic systems using neural networks," *IEEE Transactions on Neural Networks*, vol. 1, pp. 4–27, 1990.

[30] K. S. Narendra, *Neural Networks for Control*, ch. 5, pp. 115–142. MIT Press, 1990.

[31] K. S. Narendra and A. M. Annaswamy, *Stable adaptive systems.* Englewood Cliffs, NJ: Prentice-Hall, 1989.

[32] B. Widrow and S. D. Stearns, *Adaptive Signal Processing.* Englewood Cliffs: Prentice-Hall, 1985.

[33] B. Widrow, "Adaptive inverse control," in *Preprints of the 2nd IFAC workshop on Adaptive Systems in Control and Signal Processing, Lund, Sweden*, pp. 1–5, 1986.

[34] D. Psaltis, A. Sideris, and A. A. Yamamura, "A multilayered neural network controller," *IEEE Control Systems Magazine*, vol. 8, pp. 17–21, 1988.

[35] K. J. Åström and B. Wittenmark, *Adaptive Control.* Addison-Wesley, 1989.

[36] C. G. Economou, M. Morari, and B. O. Palsson, "Internal model control. 5. Extension to nonlinear systems," *Ind. Eng. Chem. Process Des. Dev.*, vol. 25, pp. 403–411, 1986.

[37] S. Lee and R. M. Kil, "Multilayer feedforward potential function network," *Neural Networks*, 1990.

[38] K. J. Hunt and D. Sbarbaro, "Neural networks for non-linear Internal Model Control," *Proc. IEE Pt. D*, vol. 138, pp. 431–438, 1991.

[39] C. A. Desoer and M. Vidyasagar, *Feedback Systems : Input-output Properties.* London: Academic Press, 1975.

[40] P. J. Werbos, *Neural Networks for Control*, ch. 2, pp. 59–65. MIT Press, 1990.

[41] E. Grant and B. Zhang, "A neural net approach to supervised learning of pole balancing," in *IEEE Int. Symposium on Intelligent Control 1989*, pp. 123–129, 1989.

[42] W. T. Miller, R. S. Sutton, and P. J. Werbos, *Neural Networks for Control.* Cambridge, Massachusetts: MIT Press, 1990.

[43] V. S. Kosikov and A. P. Kurdyukov, "Design of a nonsearching self-adjusting system for nonlinear plant," *Avtomatika i Telemekhanika*, vol. 4, pp. 58–65, 1987.

[44] C. E. Garcia and M. Morari, "Internal model control - 1. A unifying review and some new results.," *Ind. Eng. Chem. Process Des. Dev.*, vol. 21, pp. 308–323, 1982.

[45] M. Morari and E. Zafiriou, *Robust Process Control.* Englewood Cliffs: Prentice-Hall, 1989.

[46] H. Demircioglu and P. J. Gawthrop, "Continuous-time generalised predictive control," *Automatica*, vol. 27, pp. 55–74, January 1991.

[47] D. Q. Mayne and H. Michalska, "Receding horizon control of nonlinear systems," *Trans. IEEE on Automatic Control*, vol. 35, pp. 814–824, 1990.

[48] E. Ersu and H. Tolle, "A new concept for learning control inspired by brain theory," *Proc. 9th World Congress of IFAC*, vol. 7, pp. 245–250, 1984.

[49] S. S. Keerthi and E. G. Gilbert, "Moving-horizon approximations for a general class of optimal nonlinear infinite-horizon discrete-time systems," in *Proc. 20th Annual Conference Information Science and Systems, Princeton University*, pp. 301–306, 1986.

[50] D. Q. Mayne and H. Michalska, "An implementable receding horizon controller for stabilization of nonlinear systems," report IC/EE/CON/90/1, Imperial College of Science, Technology and Medicine, 1990.

[51] P. J. Gawthrop, "Hybrid self-tuning control," in *Encyclopedia of Systems and Control* (Singh, ed.), Pergamon, 1987.

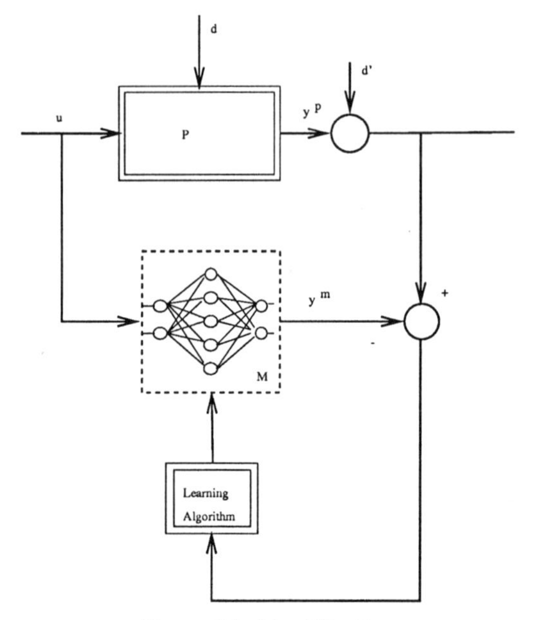

Figure 6.1: Identification

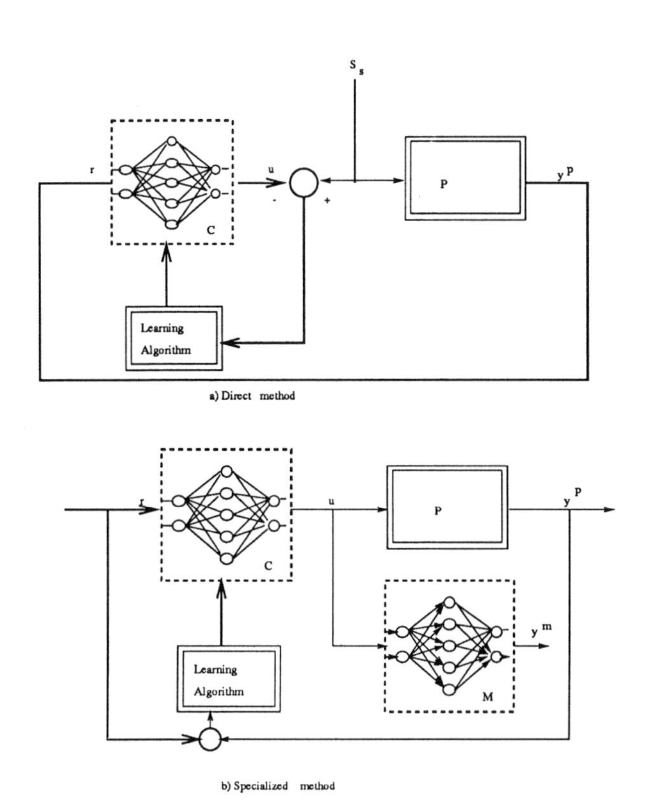

Figure 6.2: Structures for inverse identification

The object of the inverse model is to get a transformation from r → u space, where r is the desired command signal.

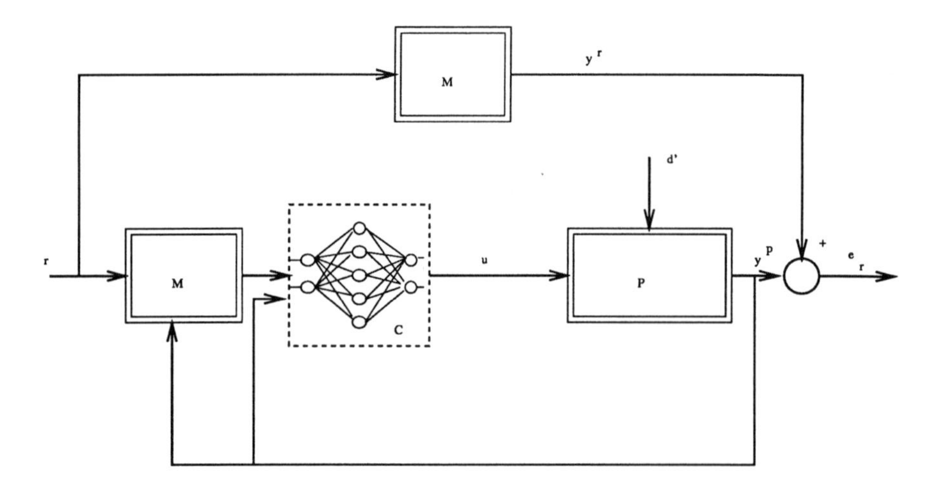

Figure 6.3: Model reference structure

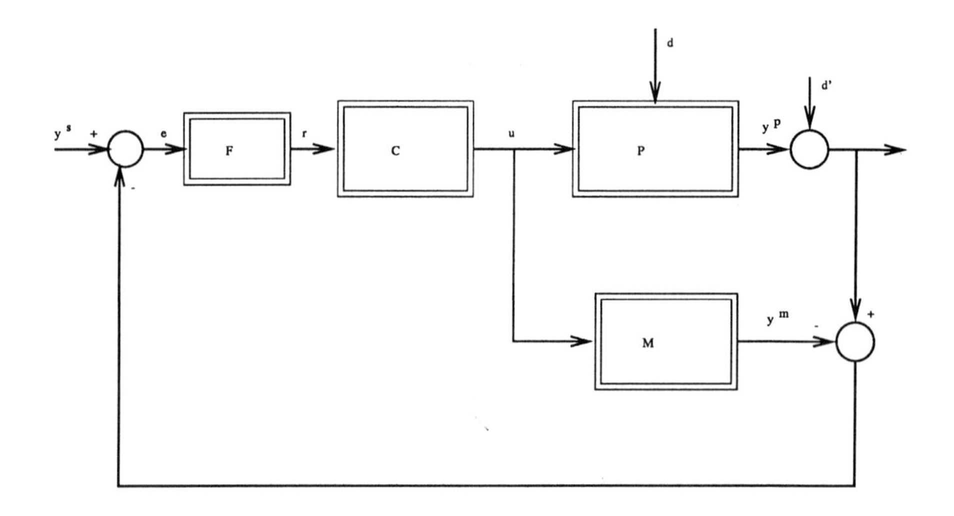

Figure 6.4: Nonlinear IMC structure

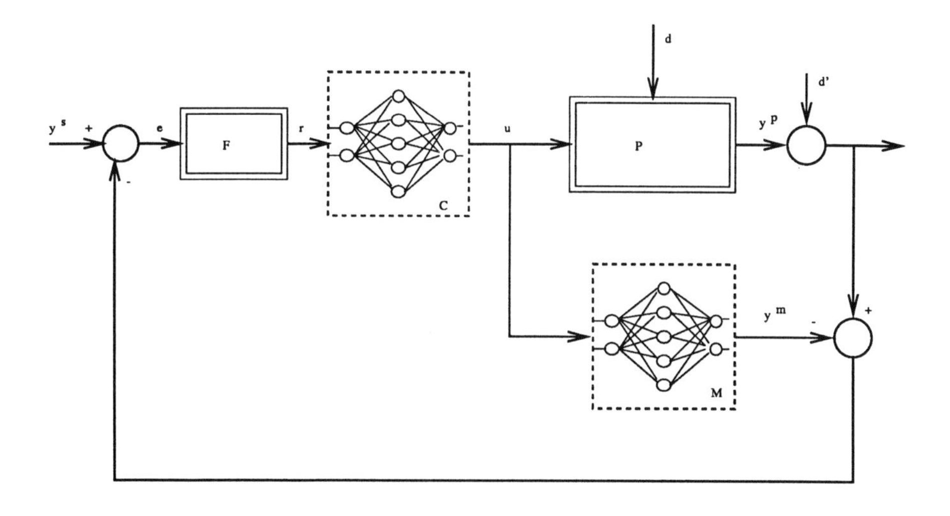

Figure 6.5: Structure for internal model control

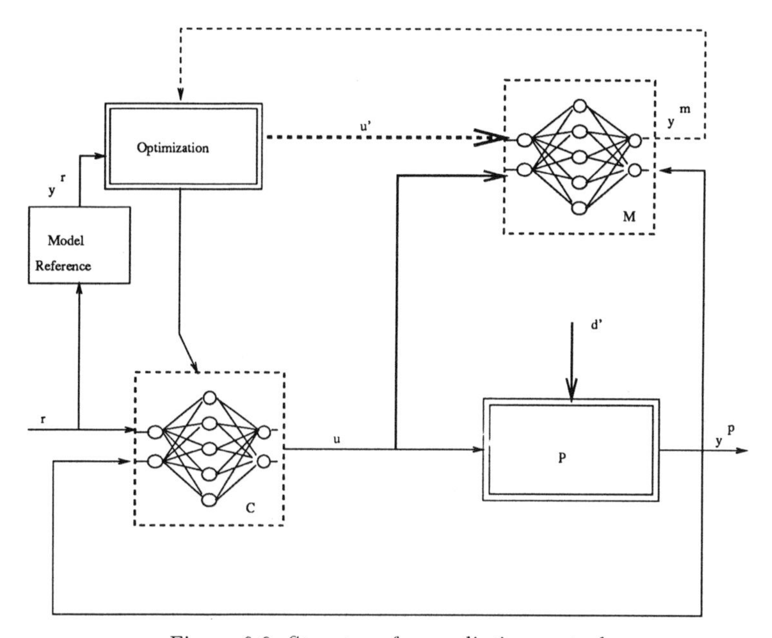

Figure 6.6: Structure for predictive control

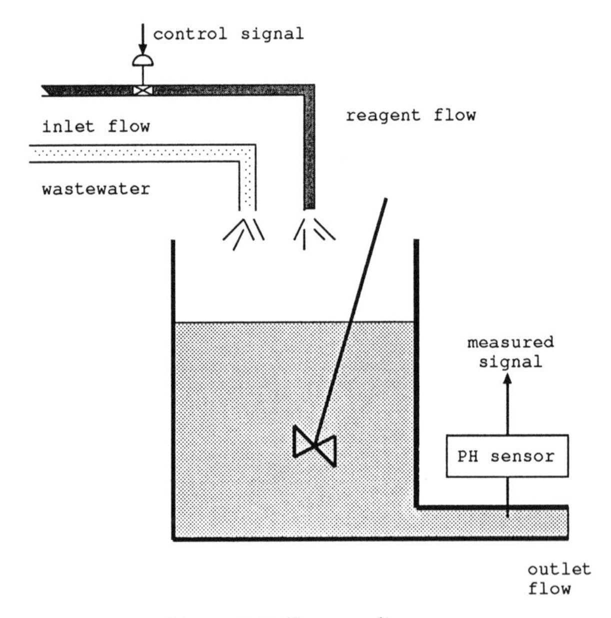

Figure 6.7: Process diagram

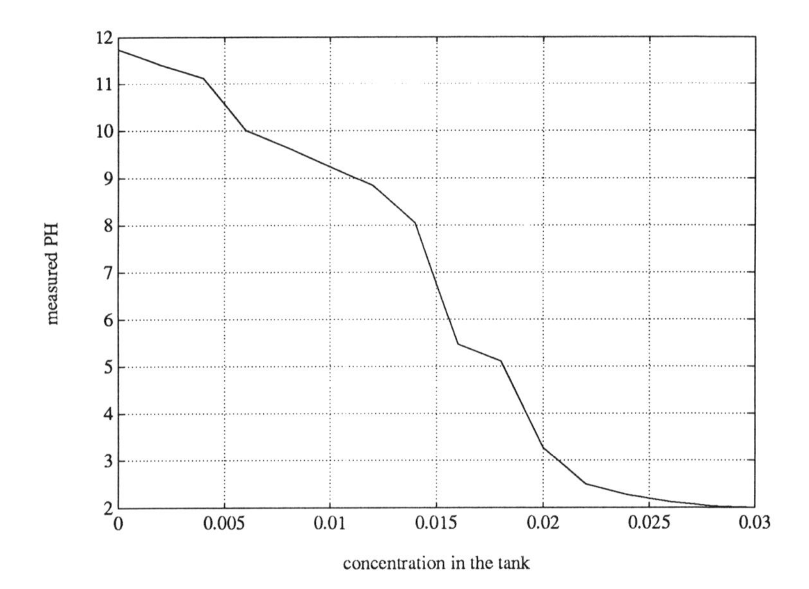

Figure 6.8: Titration curve

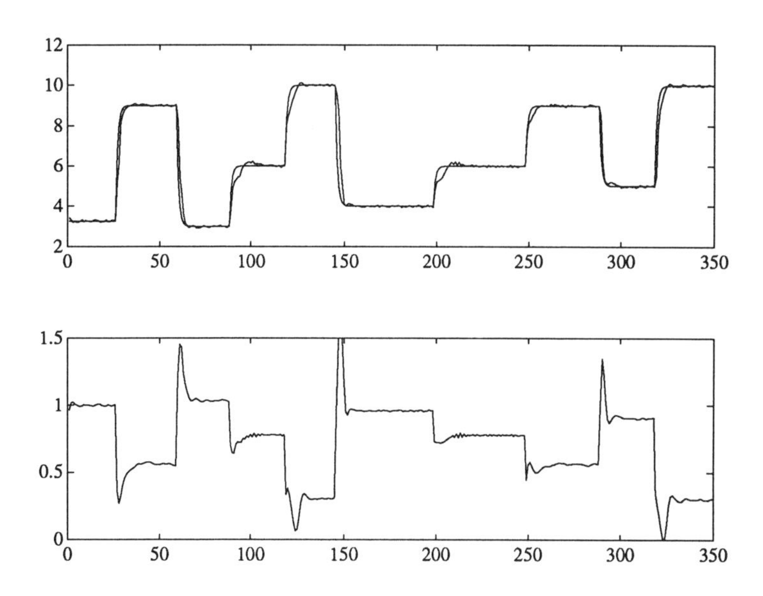

Figure 6.9: IMC control

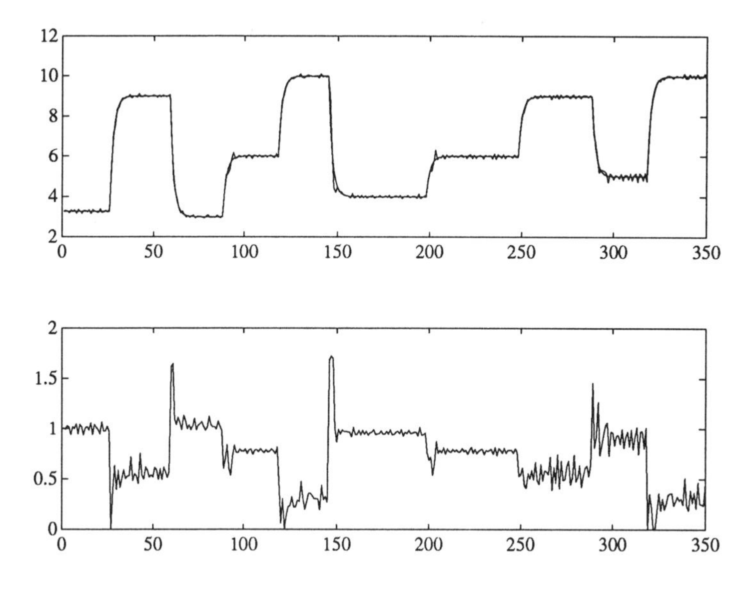

Figure 6.10: Receding horizon control: $\lambda = 0$

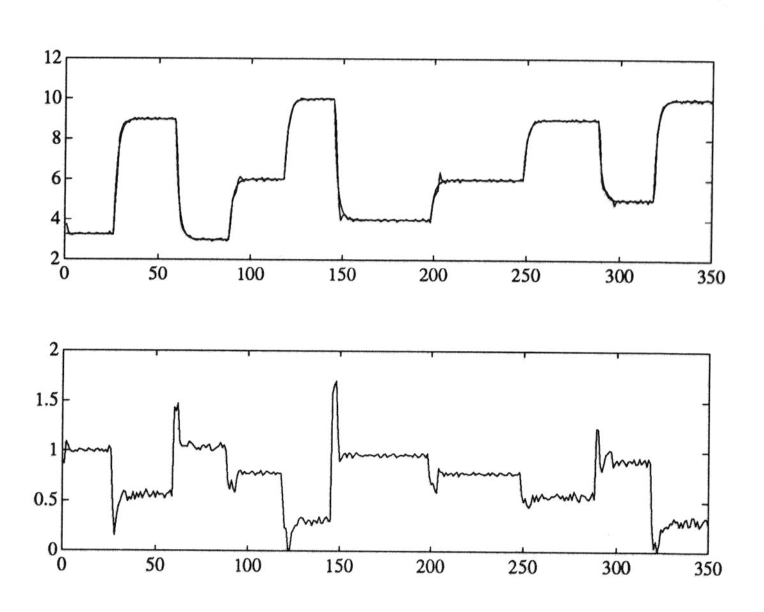

Figure 6.11: Receding horizon control: $\lambda = 1$

Chapter 7

Artificial neural networks: methodologies and applications in process control

G. A. Montague, A. J. Morris and M. J. Willis

7.1 Introduction

Hype has played a significant role in the history of artificial neural network research and application. From the initial recognition of the potential for networks to approximate non-linear functions some quite exaggerated claims have been an unwelcome accompaniment to research developments. Results from both simulation studies and real process applications indicated that even simple network architectures appeared to possess powerful non-linear process modelling capabilities that provided easy and quick practical solutions to complex problems. Optimism grew and the pragmatism emerged. In contrast, the structure of a neural network based model was also being considered generic in the sense that little prior knowledge of the process was required. Unlike Autoregressive Moving Average and Nonlinear Autoregressive Moving Average approaches, the methodology has been attributed the potential of accurately describing the behaviour of extremely complex systems. Claims that the technique offered a panacea to all modelling problems surfaced.

The interest in Artificial Neural Networks (ANN's) has arisen from the increasing attention being focussed on pattern processing techniques and their potential application to a vast range of complex and demanding real-world problems. These include, for example, image and speech processing, inexact knowledge processing, natural language processing, sensor data processing, control, forecasting and optimisation.

In the mid 1980's wide spread interest in ANN research re-emerged following a period of reduced research funding. The much wider availability and power of computing systems, together with new research studies, resulted in a far greater 'market' for the technology. The seeds were sown for claims by some that the technique provided the much sort pragmatic solutions required by industry, and by others that it provided the panacea to all complex modelling problems. The concept of an ANN as a 'black box' prescription for all modelling problems is, however, very far from reality. Indeed, selling the technology in this manner discredits what could prove to be a valuable engineering tool. If, though, ANN's only partly fulfil their projected promise they may form the basis of improved alternatives to current engineering practice. Indeed, one of the major obstacles to the widespread use of advanced modelling and control techniques, the cost of model development and validation, could be significantly eased.

Within the area of process engineering, process design and simulation; process supervision, control and estimation; process fault detection and diagnosis rely upon the effective processing of unpredictable and imprecise information. To tackle such tasks, current approaches tend to be based upon some 'model' of the process in question. The model can either be qualitative knowledge derived from experience; quantified in terms of an analytical (usually linear) process model, or a loosely integrated combination of both. Although the resultant procedures can provide acceptable solutions, there are many situations in which they are prone to failure because of the uncertainties and the nonlinearities intrinsic to many process systems. These are, however, exactly the problems that a well trained human decision process excels in solving. Thus, if ANNs fulfil their projected promise, they may form the basis of improved alternatives to current engineering practice. Indeed, applications of artificial neural networks to solve process engineering problems have already been reported.

One of the major obstacles to the widespread use of advanced modelling and control techniques is the cost of model development and validation. The utility of neural networks in providing viable process models was demonstrated in Bhat et al (1989), where the technique was used to successfully characterise two non-linear chemical systems as well as interpret biosensor data. The application of ANNs to provide cost efficient and reliable process models therefore appears to be highly promising.

In an adjunct area, the use of soft-sensor methodologies for the on-line estimation of process variables has been successfully considered at Newcastle with a number of potentially industrially attractive schemes having been proposed and validated. In some situations, techniques based on the use of an neural network models may offer significant advantages over conventional model based techniques. For instance, if such a model is sufficiently accurate, it could theoretically be used in place of an on-line analyser. Indeed, such a philosophy may be used to provide more frequent measurements than could be achieved by hardware instrumentation. This is advantageous from the control viewpoint as the feedback signals will be not be subject to measurement delays. Consequently, significant improvements in control performance can be expected. Di Massimo et al (1990), Montague et al (1991) and Willis et al (1991a,b,c) discuss these modelling techniques and demonstrate their potential applicability; Tham et al (1991), discusses soft-sensor techniques.

If an accurate network model is available, then it could obviously also be directly applicable within a model based control strategy. The particularly attractive feature is the potential to handle nonlinear systems. Psaltis et al (1988) and Narendra and Parthasarathy (1990) investigated the use of a multi-layered neural network processor for plant control. A novel learning architecture was proposed to ensure that the inputs to the plant gave the desired response.

This chapter examines the contribution that various network methodologies can make to the process modelling and control toolbox. Feedforward networks with sigmoidal activation functions, radial basis function networks and nonlinear partial least squares approaches are reviewed and discussed using data from a large industrial process. Finally the concept of dynamic networks is introduced.

7.2 System Modelling

System modelling (ie its mathematical representation) and model parameter identification are fundamental problems in process engineering where it is often required to approximate the behaviour of a real system with an appropriate mathematical model given a set of input-output data. The model structure should be selected to enable the underling system characteristics to be approximated with acceptable accuracy. For linear time-invariant systems model structure and identification problems have been well studied and the literature abounds with many useful methods, algorithms and application studies (eg Ljung, L. and Soderstrom, T., 1983). Widely used structures are the Autoregressive Moving Average (ARMA), the Autoregressive with Exogeneous Variables (ARX) and the Autoregressive Moving Average with Exogeneous Variables (ARMAX) representations.

$$y(t) = \sum_{i=1}^{a_n} a_i\, y(t-i) + \sum_{i=1}^{b_n} b_i\, u(t-i) + \sum_{i=1}^{c_n} c_i\, e(t-i) + e(t)$$

For models of the form $\mathbf{y} = \mathbf{X}\beta + e$ the least squares estimator for the regression coefficients in β is the vector $\mathbf{b}$ that satisfies:

$$\frac{\partial}{\partial \mathbf{b}} \left[(\mathbf{y} - \mathbf{Xb})'(\mathbf{y} - \mathbf{Xb}) \right] = 0$$

where $(\mathbf{y} - \mathbf{Xb})'(\mathbf{y} - \mathbf{Xb})$ represents the residual sum of squares. Ordinary Least Squares regression leads to unbiased parameter estimates if the errors e are independent of the elements of **X**. However, if the regressor variables in **X** are not independent, but exhibit a high degree of multicollinearity, the least squares estimates will have very large variance. Since the elements of **X** for the models discussed are simply lagged values of the inputs and outputs, this situation is a major problem in fitting these models. In these cases biased forms of estimation are used as alternatives to simple least squares. Biased estimation is used to attain a substantial reduction in variance with an accompanied increase in the stability of the regression coefficients. Typically, constrained least squares approaches, eg. ridge regression, are used.

In practice most systems encountered in industry are non-linear to some extent and in many applications non-linear models are required to provide acceptable representations. Non-linear system identification is, however, much more complex and difficult although the Nonlinear Autoregressive Moving Average with Exogeneous Variables (NARMAX) description has been shown to provide a very useful unified representation for a wide class of non-linear systems:

$$y(t) = \mathbf{f}[\, y(t-1),.., y(t-n); u(t-1),.., u(t-n); e(t-1),..,e(t-n) \,] + e(t)$$

Here **f[.]** is a nonlinear function which is rarely known *a priori* and can be very complicated. Efficient parameter identification procedures are particularly important with non-linear systems so that the resulting models are not over parameterised. The work of Billings and colleagues (eg Leontaritis and Billings, 1985) stand out in the contributions made in this area.

The problem of identifying, or estimating, a model structure and its associated parameters can be related to the problem of learning a mapping between a known input and output space. A classical framework for this problem can be found in *approximation theory*. Almost all approximation, or identification, schemes can be mapped into, ie. expressed as, a network. For example, the well known ARMAX model can be represented as a single layer network with inputs comprising of lagged system input-output data and prediction errors. In this context a network can be viewed as a function represented by the conjunction of a number of *basic* functions. [A set of *basis* vectors for a space Ω is a subset of vectors in Ω which (a) spans the space Ω, and (b) is a linearly independent set. Alternatively, a set of basis vectors is a set consisting of the minimum number of vectors required to span the space Ω. Basis vectors can be considered as the generalisation of the notion of coordinate vectors in two and three dimensioned space].

The Artificial Neural Network Model

In developing a model which is representative of a systems behaviour, it is the topology of the network, together with the neuron processing function, which determine the accuracy and degree of the representation. A literature search reveals well over 50 different types of network architecture and a number of different neuron processing functions. This paper concentrates on one specific topology which has probably been the most prevalent in ANN studies, the feedforward (often termed backpropagation) network.

The power of the feedforward approach has been demonstrated by a number of workers (eg, Hecht-Nielson, 1989). A number of papers which have indicated that a feedforward network has the potential to approximate any non-linear function. Cybenco (1989) and Wang et al (1991) have shown that the two-layered feedforward network can uniformly approximate any continuous function to an arbitrary degree of exactness - providing that the hidden layer(s) contains a sufficient number of nodes.

The ability of the network to approximate non-linear functions is dependent upon the presence of hidden layers. Without these only linear combinations of functions can be fitted. The number of nodes in the hidden layer(s) can be as small or large as required. It is related to the complexity of the system being modelled and to the resolution of the data fit.

Approximation theory addresses the problems of interpolating or approximating a continuous multivariate function, **f**(x), by a selected function, **F**(x,w), having a fixed number of parameters, w. Given a specific function **F** the problem reduces to determining (or learning) the set of parameters w that provides the best possible approximation to **f** (according to *a priori* chosen criteria), when exposed to (or trained on) an 'example' data set. The problem of choosing an appropriate approximating function **F** is known as the representation problem.

The basic feedforward network performs a non-linear transformation of input data in order to approximate output data. The number of input and output nodes is determined by the nature of the modelling problem being tackled, the method of input data representation chosen, and the form of network output required. The input layer to the network does not perform processing but merely acts as a means by which scaled data is introduced to the network. The data from the input neurons is propagated through the network via the interconnections. The interconnections within the network are such that every neuron in a layer is connected to every neuron in adjacent layers. It is the hidden layer structures which essentially define the topology of a feedforward network. Each interconnection has associated with it a scalar weight which acts to modify the strength of the signal. The neurons within the hidden layer perform two tasks; they sum the weighted inputs to the neuron and then pass the resulting summation through a non-linear processing, or activation, function. In addition to the weighted inputs to the neuron, a bias is included in order to shift the space of the non-linearity.

For example, if the information from the ith neuron in the $j\text{-}1^{th}$ layer, to the kth neuron in the jth layer is $I_{j-1,i}$, then the total input to the kth neuron in the jth layer is given by:

$$S_{j,k} = b_{j,k} + \sum_{i=1}^{n} w_{j-1,i,k}\, O_{j-1,i}$$

where $b_{j,k}$ is a bias term which is associated with each interconnection.

The output of each node is obtained by passing the weighted sum, $S_{j,k}$, through a nonlinear operator. The most widely applied non-linearity is the sigmoidal function in the interval (0,+1). In situations where both positive and negative outputs are required the function can be re-scaled into the interval (-1,+1). The sigmoidal function has the mathematical description:

$$O_{j,k} = 1/[1 + \exp(-S_{j,k})]$$

Although this function has been widely adopted, in principle, any function with a bounded derivative could be employed (Rumelhart et al, 1987). It is, however, interesting to note that a sigmoidal nonlinearity has also been observed in human neuron behaviour, (Holden, 1976). Within an ANN, this function provides the network with the ability to represent nonlinear relationships. Additionally, note that the magnitude of bias term effectively determines the co-ordinate space of the nonlinearity. This implies that the network is also capable of characterising the structure of the nonlinearities: a highly desirable feature.

Other alternative nonlinear activation functions are for example the monotonic forms such as hyperbolic tangent, simple quadratic, etc., and those based on Radial Basis functions such as, for example, Gaussian distributions.

7.3 Feedforward Network Training and Validation

To develop a process model using the neural network approach, the topology of the network must first be declared. The determination of the number of nodes in the hidden layers for feedforward networks tends, in the main, to be based upon heuristic approaches, although methodologies are now being developed which assist in hidden layer specification based upon rank degeneracy of the matrix formed from the hidden layer(s) output space (Wang et al., 1991, 1992).

Process modelling using ANNs is similar to identifying the coefficients of a parametric model of specified order. Loosely speaking, specifying the topology of an ANN is similar to specifying the 'order' of the process model. For a given topology, the magnitudes of the weights define the characteristics of the network. However, unlike conventional parametric model forms, which have an *a priori* assigned structure, the weights of an ANN also define the structural properties of the model.

The problem of determining the network parameters (weights) can be considered essentially a non-linear optimisation task. The *simplest* optimisation technique makes use of the Jacobian of the objective function to determine the search direction. A 'learning' rate term which influences the rate of weight adjustment (Rumelhart and McClelland, 1987) is used as the basis for the back-error propagation algorithm, which is a distributed gradient descent technique. In order to train the network with this approach, a representative process input data set is presented to the network. At each time instance, the input set is propagated through the network to give a prediction of output. The error in prediction is then used to update the weights based upon gradient information, in order to drive the cost function to a minimum. The error in prediction is thus in a sense back-propagated through the network to update the weights. Such an approach to training is termed supervised learning since at any time instant both input and output data is available.

A numerical search technique is used to determine the weights. The objective function for the optimisation is written as:

$$V(\Theta,t) = 1/2 \sum E(\Theta,t)^2$$

where 'Θ' is a vector of network weights, 'E' is the output prediction error and 't' is time. The search direction can be generalised as:

$$\Theta_{t+1} = \Theta_t + \delta_t S \nabla V(\Theta,t)$$

where δ_t is the 'learn' rate which influences the rate of weight adjustments, and S is the identity matrix. This was used by Rumelhart and McClelland (1987) as the basis for their 'back-error propagation' algorithm, which is a distributed gradient descent technique. The popularity of this learning technique can be judged by the commonly adopted, somewhat misleading, description of feedforward nets as backpropagation nets. There are, however, a number of problems with the backpropagation approach. In common with other descent algorithms, difficulty arises when the search approaches the minima. If the surface is relatively flat in this region then the search becomes inefficient. Therefore, in most neural network applications, a 'momentum' term is added. The current change in weight is forced to be dependent upon the previous weight change. In this approach, weights in the jth layer are adjusted by making use of locally available information, and a quantity which is 'back-propagated' from neurons in the $j+1^{th}$ layer.

$$\Theta_{t+1} = \Theta_t + \delta_t S \nabla V(\Theta,t) + \beta(\Theta_t - \Theta_{t-1}) \quad 0 < \beta < 1$$

Here, 'ß' is a factor which is used to influence the degree of this dependence. Although this modification does yield improved performances, in training networks where there are numerous weights, gradient methods have to perform exhaustive searches and are also rather prone to failure.

Although this modification does yield improved performances, gradient techniques can require significant convergence times in large dimensioned problems. Furthermore, in adopting a down-hill search technique, the question arises as to whether the minimum is local or global. Although it could be argued that a momentum term may take the solution over a local minima, global optimality is not assured.

A more appealing method is that of Conjugate Gradients, Leonard and Kramer (1990). Although quasi-Newton methods are usually more rapidly convergent and more robust than conjugate gradient methods, they require significantly more storage. A Newton-like algorithm, which includes second derivative information, would treat the S matrix, in the optimisation procedure discussed above, as the inverse of the

Hessian. An advantage of the conjugate gradient method, however, is that it relinquishes the need for second derivatives of the objective function whilst retaining convergence properties of second order techniques. A conjugate gradient methodology is thus a well established contender for problems with a large number of techniques. A conjugate gradient technique is a well established contender for problems with a large number of variables such as the training of an artificial neural network.

The basic philosophy is to generate a conjugate direction as a linear combination of the current steepest descent direction and the previous search direction. With this technique minimisation is initiated as with steepest descent. After this iteration a new direction of search is required. This direction is chosen so that it is conjugate with the initial search direction. Minimisation then proceeds in the newly defined direction. It should be noted, however, that when this technique is applied to non quadratic functions, the exact minimum will not be found in a finite number of steps. Practical experience suggests that resetting the algorithm to the steepest descent direction every n iterations (where n is the number of network weights) is superior to the repeated use of the conjugate gradient method.

An alternative approach (Bremermann and Anderson, 1989), attempts to avoid the solution becoming locked into a local minima. Postulating that weight adjustments occur in a random manner and that weight changes follow a multivariate zero mean Gaussian distribution, the algorithm adjusts weights by adding Gaussian distributed random values to old weights. The new weights are accepted if the resulting prediction error is smaller than that from the previous set of weights. During minimisation the allowable variance of the increments can be adjusted to assist network convergence and aid in the avoidance of local minima. The algorithm is of the graded learning type. Network outputs do not need to be available at every time instant, merely a measure of quality of fit needs to be given periodically.

The chemotaxis algorithm can be summarised as follows:

Step I Initialise weights with small random values

Step II Present the inputs, and propagate data forward to obtain the predicted outputs

Step III Determine the cost of the objective function, E_1, over the whole data set.

Step IV Generate a Gaussian distributed random vector.

Step V Increment the weights with random vector.

Step VI Calculate the objective function, E_2, based on the new weights.

Step VII If E_2 is smaller than E_1, then retain the modified weights, set E_1 equal to E_2, and go to Step V. If E_2 is larger than E_1, then goto Step IV.

It is noted that during the minimisation, the allowable variance of the increments may be adjusted to assist network convergence and aid in the avoidance of local minima.

In network training one method adopted is to split the data, randomly, into training and test data sets. If the network approximation is adequate then the squared error between the training data outputs and network predicted outputs should be 'relatively' small and, more importantly, should be uncorrelated with all combinations of past inputs and outputs. Comparison of the average error achieved on the training set with that achieved on the test data set can be used to indicate the adequacy of the model (network). For example, if the test data set error is significantly larger then an over-dimensioned network is indicated.

A number of model validity tests for non-linear model identification procedures have been developed, for example the statistical chi-squared test (Leontaritis and Billings, 1987), the Final Prediction Error Criterion (Akaike, 1974), the Information Theoretic Criterion (AIC) (Akaike, 1974) and the Predicted Squared Error criterion (Barron, 1984). The PSE criteria, although originally developed for linear systems, can be applied to feedforward nets providing that they can be approximated by a linear model.

$$\text{Final Prediction Error (FPE)} = (E/2N)\,(N + N_W)/(N - N_W)$$
$$\text{Information Theoretic Criterion (AIC)} = \ln(E/2N) + 2\,N_W/N$$
$$\text{Predicted Squared Error (PSE)} = E/2N + 2*(\sigma^2)N_W/N$$

where σ^2 is the prior estimate of the true error variance and independent of the model being considered

It can be seen that these tests make use of functions that strike a balance between the accuracy of model fit (average squared error over N data points, E/2N) and an overfit penalty which is a function of the number of adjustable parameters or weights used (N_W). Because of this, minimisation of these test functions results in networks (models) that are neither under nor over dimensionalised. A procedure involving 'train - test - validate' experiments is used with different network dimensions to minimise a selected validation function.

Final validation of the identified network model should be achieved by checking its predictive qualities for both one-step-ahead and multi-step-ahead predictions. In addition it is also informative to plot the residuals.

One approach to the assessment of network dimensionality required to model a particular data set is to adopt a structural identification method used with ARMA models. Here an information matrix constructed from input-output data sequences, and in product moment form used to determine the 'best' model order by searching for rank degeneracy. In a least squares context, the uniqueness of the parameter estimates is determined by the rank of the matrix. If the matrix has full rank, then the estimated parameter vector has a distinct solution. To assist in interpretation the determinant can be normalised using the ratio of consecutive determinants. It is worth noting, however, that data sampling variabilities and computational errors could make this approach problematical (Hankle matrices constructed from real plant data, and used to assess minimal realisation, tend to be always full rank).

With a network, an 'information matrix' is constructed for each hidden layer, using the output vectors from each node in the hidden layer, created as a result of presenting the complete data set to the network. It is then argued that, providing each node output vector can be approximated by a linear vector, any hidden layer is considered 'under specified' if the vectors in the matrix are linearly independent. Conversely, the layer is considered 'over-specified' if the vectors are linearly dependent. The 'best' representation of the information in the layer is provided at the point when the vectors just become linearly dependent. Other solutions will be arbitrary points in the kernel space. This method provides a maximum sub-space for the nonlinear function fitting (Wang et al, 1992).

Neural network model identification requires that the input data sequence must, at least, satisfy all the well known conditions associated with MIMO parameter and structural identification - i.e. all the system 'modes' must be excited. However, persistent excitation alone is not sufficient. Random excitation is required with a magnitude covering the whole dynamic range and density that is sufficient to encapsulate the whole input domain of interest. The excitation required for good identification of an adequate model will be closely related to the properties of the system being modelled. This in turn is related to the distribution of the training data set(s).

7.4 Radial Basis Function Networks

Recently there has been a renewed interest in Radial Basis Functions (RBF) within the engineering community. RBF's are a traditional and powerful technique for interpolation in multidimensional space (Micchelli, 1986; Powell, 1985; Powell, 1987). A generalised form of the RBF has found wide application in areas such as image processing, signal processing, control engineering, etc. An RBF expansion with n-inputs and a scalar output implements a mapping according to:

$$\mathbf{f(x)} = \beta_0 + \sum_{i=1}^{n} \beta_i \phi(||\mathbf{x} - \mathbf{c}_i||)$$

where

β_i are the parameters or weights of $\mathbf{c}_i$,
$\mathbf{c}_i$ are the RBF centres and n is the number of centres.

The functional form $\phi(.)$ is pre-selected with the centres c_i being some fixed points in n-dimensional space appropriately spanning the input domain.

It is interesting to observe that there is a strong relationship between radial basis function approaches and neural network representations. Indeed, if the RBF centres are not predetermined but are considered to be adjustable parameters (cf the w parameters in the approximation function approach), then a two layered feedforward network can be obtained. Since 'centres' can be considered to be adjustable within the network, such a representation provides many more degrees of freedom allowing task-oriented network design. Such flexibility, although very powerful in dealing with complex data, does not come cheaply. The network structure is now highly non-linear in the parameters. The radial basis function approach is, however, only one of a large class of possible function approximators.

The RBF representation can be implemented in the form of a two-layered network. For a given set of centres, the first layer performs a fixed nonlinear transformation which maps the input space onto a new space.
Each term $\phi(|| \mathbf{x} - \mathbf{c}_i ||)$ forms the activation function in a unit of the hidden layer. The output layer then implements a linear combination on this new space. With the RBF centres regarded as adjustable parameters a network structure in feedforward form results.
The most popular choice for $\phi(.)$ is the Gaussian form and in this case the activation function in any hidden unit (h) becomes:

$$\mathbf{O}_h = \exp\left[- \sum_{i=1}^{n} (\mathbf{x}_i - \mathbf{c}_{h,i})^2 / \sigma_i^2\right]$$

where each Gaussian is characterised by:
the function centre, $\mathbf{c}_{h,i}$
the function width, σ_i.
The function height is determined by the weights in the output layer.

The output layer provides a biased linear combination of the hidden layer outputs:

$$\mathbf{Y}_k = \mathbf{B}_k + \sum_{i=1}^{n} \mathbf{W}_{k,i}\, \mathbf{O}_i$$

The nonlinear mapping (or model fit) is composed of a linear combination of Gaussian functions (non-monotonic), instead of Sigmoids (monotonic). The surface (in two dimensions) generated by this form of activation function appears like a series of hills. The "heights" of these "hills" are varied by adjustment of the scalar weights in the second (output) layer. If a sufficient number of hidden layer neurons are provided then the nonlinear function can be well approximated by varying the centres $\mathbf{c}_i$, the "width" scaling parameter σ and the output layer weights. The work of Chen, Billings and Grant (1990) and Leonard, Kramer and Ungar (1991) make significant contributions to understanding and applications in this important area.

7.4.1 Radial Basis Function Network Training and Validation

The Radial Basis Function Networks have three layers, namely the input layer, the hidden layer with non-linear activation functions and the linear output layer. Only the hidden layer and output layer need to be trained though. This is because the input layer acts as a fan-out only for the input vectors and therefore all the weights on the links between the input layer and the hidden layer are fixed to 1.0.

Training of the hidden layer consists of positioning the cluster centres in the input space spanned by the inputs and determining the width of each of the clusters. For the positioning of the centres K-Means clustering with a ordinary euclidean distance measure of the input vectors, scaled to have zero mean and unit variance, is used. For the determination of the width of each of the clusters the p-nearest clusters centres are located and the distances between the current cluster centre and those p-nearest clusters centres is then combined into the width parameter sigma. It should be noted that at this time the only data used is the training input data. No reference is being made as yet to any training output data.

Once the hidden layer is configured, all inputs are fed through the hidden layer and stored in a matrix. In order to determine the weights of the connections between the hidden layer and the output layer, generalised least squares minimisation is used, using the matrix of hidden layer outputs and the training output data. The generalised least squares method is implemented with the Singular Value Decomposition algorithm, which guarantees the optimum solution.

The two parameters that control the networks estimation capabilities are the number of clusters and the number of neighbouring clusters that should be used to determine the width of each of the clusters. Network configuration comes down two choosing the optimal combination for these two parameters, where optimal is meant in the minimum predictive squared error sense. In order to obtain an estimate of the Predictive Squared Error the S-Fold cross-validation procedure is used. This method is derived from the statistical resampling theory and has been proven to give an unbiased estimate of the predictive squared error. In practice the training data set is partioned into a large and a small subset. The network is trained on the large subset and the small subset is then propagated through the network and the squared error is calculated, after which the original training set is partioned differently and the process is repeated. It continues until all elements of the original training set have been used once for testing. After this procedure has finished, the squared errors are combined to give an unbiased estimate of the predictive squared error. The configuration with the minimum estimated predictive squared error is the optimal configuration.
This method of cross-validation has the side-effect of now having available a measure of the variance of each of the outputs of the network, which in turn can be used to generate a 90% confidence interval round an estimate of the network, using the student-T distribution. Because of the distance criterion that is built into a Radial Basis Function, it is also possible to derive a measure of reliability, based on the activation of each of the clusters, the width of the clusters and the data density. With this measure it is possible to check whether the current input to the network is near the training set or not. A high reliability indicates closeness, while a low reliability

indicates that the input is "far" away from the training set and therefore that the network is actually extrapolating.

7.5 Complementary Statistical Methods

Principal Component Analysis (PCA) and Partial Least Squares, or Projection to Latent structures, (PLS) are multivariable statistical techniques which consider all the noisy and highly correlated measurements made on a process, but project the information down onto low dimensional subspaces which contain all the relevant information about the process. A general description of PCA can be found in many textbooks (eg Anderson, 1984; Review - Wold, 1987).

The development of PLS is more recent and current interest in the methodology was sparked by Wold (1982) with his analysis of socio economic problems. PLS was later revised and applied in the Chemometrics field by Wold and colleagues. Most recently, PLS has been applied to process identification (Ricker, 1988; Wise and Ricker, 1990).
The methods address the problems of:

a) being able to deal with collinear data of high dimension in both the independent (**X**) and dependent (**Y**) variables.
b) being able to substantially reduce the dimension of the problem.
c) being able to provide good predictions of (**Y**) when both process (**X**) and quality (**Y**) variables are present.

7.5.1 Principal Component Analysis (PCA)

PCA is a procedure used to explain the variance in a single data matrix (**X**). PCA calculates a vector, called the first principal component, which describes the direction of the greatest variability. It is calculated as the least squares fit of a line through the data in the space spanned by **X**, or as that linear combination of the columns of **X** ($\mathbf{Xp_1}$), given by the first eigenvector of $\mathbf{X^TX}(\mathbf{p_1})$. The second principal component is orthogonal to the first principal component and explains the greatest amount of the remaining variability. It is obtained by fitting a least squares line through the residuals left after fitting the first principal component, or as the linear combination of the columns of **X** ($\mathbf{Xp_2}$), given by the second eigenvalue of $\mathbf{X^TX}$ ($\mathbf{p_2}$).
This approach is proceeded until k principal components are obtained. Thus the axes of **X** have been rotated to a new orthogonal basis. For large data sets it is often found that the first A principal components, where A<<k, will 'explain' most of the variation in the data matrix **X**. By stopping at this point and summarising the information in **X** using new variables ($\mathbf{t_i}$), defined as those linear combinations of the x's ($t^i = Xp^i$, i=1,2,...A), given by the first A principal components, the dimensionality of the data space has been reduced from k to A. In geometric terms this is equivalent to approximating the k dimensional observation space by the projections of the observations onto a much smaller A dimensional hyperspace.

The principal components define the plane of greatest variability, and the loading vectors associated with these principal components define the location of the plane in terms of the original variables. Each observation is located on this plane via its scores (a name adopted from the social sciences). The score is the distance from the origin of the plane (**x**) along each principal component, and is calculated as the product of the loading vector and the observation. The perpendicular distance from each observation to the plane is the residual for that observation. Algebraically the X matrix has been approximated by:

$$\mathbf{X} = t_1\, p_1' + t_2\, p_2' + ... + t_A\, p_A' + \mathbf{E}$$

where **E** is a residual matrix. Ideally the dimension A is chosen such that there is no significant process information left in **E**. That is, **E** should represent random error. Thus adding a further, (A+1), principal component would only be fitting some of this random error and thereby increasing the prediction error of the principal component model.

In PCA it is very important that the data is scaled correctly since the variance contribution of a particular **x** to the total variation in **X** is dependent upon its units of measurement. In general each variable should be scaled relative to the others in terms of its relative importance. Often scaling is achieved to a common unit variance, although care must be taken not to scale up the variance of variables that are almost constant in order not to disturb the natural relationships amongst variables of the same type.

7.5.2 Partial Least Squares (PLS)

Often one group of variables (**Y**) are of greater importance, eg product quality variables, and should be included in the monitoring problem. Unfortunately these variables are usually measured much less frequently than the normal process variables. A major problem is to use the information contained in the process variables (**X**) to predict, monitor and detect changes in the output variables (**Y**). Multiple linear regression (MLR) is the most common method for developing multivariable statistical models. Unfortunately, it is well established that MLR may have severe problems dealing with large sets of collinear data, leading to very imprecise parameter estimates and poor predictions.

The PLS algorithm does not attempt to calculate a regression relationship directly and thus avoids singularity problems associated with the inversion of **X'X**. Linear PLS can be regarded as a model consisting of two relations, one known as the outer relation - itself having two parts, the other as the inner relation. PLS builds the relationship sequentially, one dimension at a time. Each PLS dimension defines two new latent variables, t_i in **X** and in u_i in **Y**, such that the correlation between t_i and u_i is maximised.

The final PLS representations of the **X** and **Y** spaces are given by:

$$\mathbf{X} = \mathbf{t}\,\mathbf{p}' + \mathbf{E}$$
$$\mathbf{Y} = \mathbf{u}\,\mathbf{q}' + \mathbf{F}$$

These two relationships can be converted into an equivalent regression relationship between the original **X** and **Y** matrices, that is $\mathbf{Y} = \mathbf{X}\boldsymbol{\beta}$. They also formulate a partial outer model. An inner model $\mathbf{u} = \mathbf{f}(t)$ is then used to relate the score vectors. For linear PLS this becomes $\mathbf{u} = \mathbf{bt} + \mathbf{h}$.

If PLS is used on a data set of uncorrelated variables or if the number of transformed variables equals the original number of variables, then the results obtained are equivalent to multiple linear regression. The advantages of PLS are most evident in the computations and predictions in systems when the number of independent underlying latent variables in the **X** and **Y** spaces is much smaller than the dimension of these matrices.

Often, when the inner relation between the scores is examined, the data does not suggest a linear fit. Discrepancies between the scores data and the fitted straight line result in modelling error. This could be reduced by fitting a curve which is more appropriate to the data, ie a nonlinear inner model. For nonlinear PLS approaches, $\mathbf{f}(t)$ takes on a simple nonlinear form, $\mathbf{u} = N(t) + \mathbf{h}$. Typical relationships have include second or third order polynomial forms.

7.5.3 Principal Component Analysis, Partial Least Squares and Neural Networks

Artificial neural networks are not a new modelling tool that replaces other well understood methods. Rather they are an additional tool and complement existing techniques. The power of a network approach can be significantly enhanced by using statistical analysis techniques to assess and transform the raw data prior to network processing. Principal component analysis methods are particularly useful as a method of pre-processing process data to initialise neural network optimisation routines.
An interesting and potentially very powerful concept is to embed a neural network model within a PLS inner model. Recent work has addressed this, Qin and McAvoy (1991), and has produced a new approach to nonlinear process characterisation. As in linear PLS, the number of factors used in the final model needs careful consideration. Often the small factors are considered to be noise and are discarded. Whilst using all the factors should produce the best fit of the training data, this may not provide for the best prediction model for unseen data. The number of factors should be chosen by assessment of the test data prediction error.

7.6 Studies of Artificial Neural Network Approaches

A major industrial bioprocess has been selected to demonstrate the performances of the three neural network approaches discussed.

The industrial fermentation for the production of penicillin is operated as a fed-batch process in which two distinct operating regimes can be identified. In the early stages of the fermentation the process is operated to produce large quantities of biomass, predominantly utilising the substrate present in the initial batched media. Towards the end of this phase the feed additions being made to the fermenter are increased as the initial substrate becomes exhausted. During the second phase of the fermentation the substrate additions are maintained at a rate which keeps the substrate concentration in the broth at a low level. As a consequence of the low substrate concentration and the resulting low growth rate, penicillin is produced in large quantities by the organism. In order to maximise the yield of penicillin it has been observed that the growth rate should be maintained above a pre-determined minimum value. The higher the growth rate is above this minimum constraint the lower the yield of penicillin. If the growth rate falls below this minimum value then irreparable damage to penicillin productivity results.

The present operating regime of off-line analysis to determine the status of the fermentation results in a conservative feeding strategy, since samples are taken relatively infrequently. It is therefore highly desirable to gain some insight as to the fermentation behaviour at a higher frequency so as to be able to operate closer to the minimum growth constraint. To achieve this goal requires a move from off-line analysis to on-line measurement or estimation. It is for this reason that state observers are being developed to provide this information and utilise it in a control scheme in order to improve fermentation operation.

7.6.1 Artificial Neural Networks in Estimation

A problem of present and increasing industrial interest where artificial neural networks may be of benefit is in providing a means by which to improve the quality of on-line information available to plant operators for plant control and optimisation. Major problems exist in the chemical and biochemical industries (paralleled with the food processing industries) concerned with the on-line estimation of parameters and variables that quantify process behaviour. In a nutshell, the problem is that the key 'quality' variables cannot be measured at a rate which enables their effective regulation. This can be due to limited analyser cycle times or a reliance upon off-line laboratory assays. An obvious solution to such problems could be realised by the use of a model along with secondary process measurements, to infer product quality variables (at the rate at which the secondary variables are available) that are either very costly or impossible to measure on-line. Hence, if the relationship between quality measurements and on-line process variables can be captured then the resulting model can be utilised within a control scheme to enhance process regulation. The concept is known as inferential estimation. Historically, with varying degrees of success, linear models, adaptive models and process specific mechanistic models have been used to perform this task. It is suggested however, that the use of a neural network model, because it is not process specific and has the ability to capture nonlinear process characteristics may be beneficial for these applications. Indeed, in this area results from industrial evaluations have been promising (eg. see Willis et al, 1991b). Whilst such schemes operating in 'open-loop' can be used to assist process operators, it should be noted that with the availability of fast and accurate product quality estimates, the possibility of closed loop **inferential control** becomes feasible. Here the inferred estimates of the controlled output are used for feedback control. The effective elimination of a time delay caused by the use of an on-line analyser or the need to perform off-line analysis affords the opportunity of tight product control via the use of standard industrial controllers. Consequently reduction in product variability caused by process disturbances, and hence a reduction of off-specification product can be achieved.

Many industrial systems subject to measurement limitations could be considered for demonstration purposes. However, the complexity of the penicillin fermentation process is an good example to highlight the possible benefits of applying neural network modelling techniques. In this process biomass levels inside a fermenter need to be regulated to achieve good performance. Unfortunately, the only method available to determine biomass is by off-line laboratory analysis. The delay induced by this sampling procedure and the frequency at which samples can be taken reduce the effectiveness of control. A complex non-linear relationship is known to exist between on-line measured variables, such as off-gas concentrations, and biomass levels in the fermenter. Thus the objective was to model this relationship using the neural network modelling procedure. A series of experimental trials were performed and the results were utilised to provide training data linking off-gas to biomass levels. The current on-line measurement of carbon dioxide evolution rate in the off-gas (CER) was used as one of the inputs to the neural network. Additionally, since the characteristics of the fermentation are also a function of time, the batch time was also considered a pertinent input. Neural networks have been trained to capture the complex fermentation relationship and provide estimates of biomass at a much increased frequency.

7.6.2 Results and Discussion

Three data sets were used to demonstrate the ability of each particular network to represent the process under consideration. Two data sets were used for training purpose whilst the third one highlight their aptitude to generalise. However, for industrial confidentiality reasons, the actual level of biomass achieved is not shown.

a) <u>The RBF Neural Network</u>

As described in section 7.4.1 the Radial Basis Function networks are configured using S-Fold cross-validation in order to minimise the predictive squared error. It is therefore not surprising to find that these networks manage to reproduce the behaviour of the training set almost exactly, as is shown in figures 1 and 2. However, when a test data set is processed, the network does not manage to reproduce the

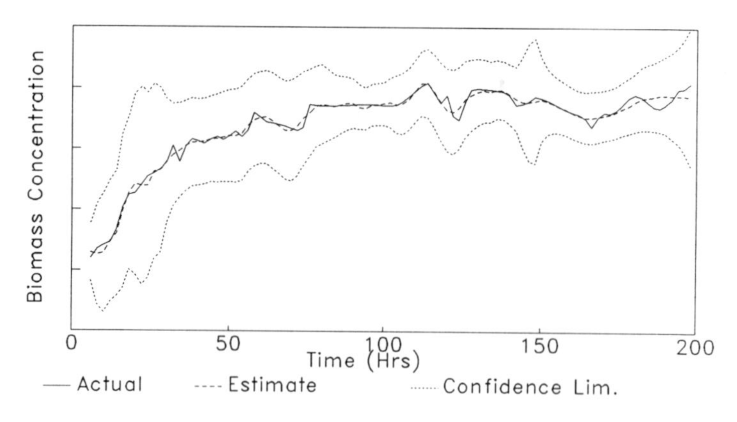

Figure 1. RBF: Training set #1

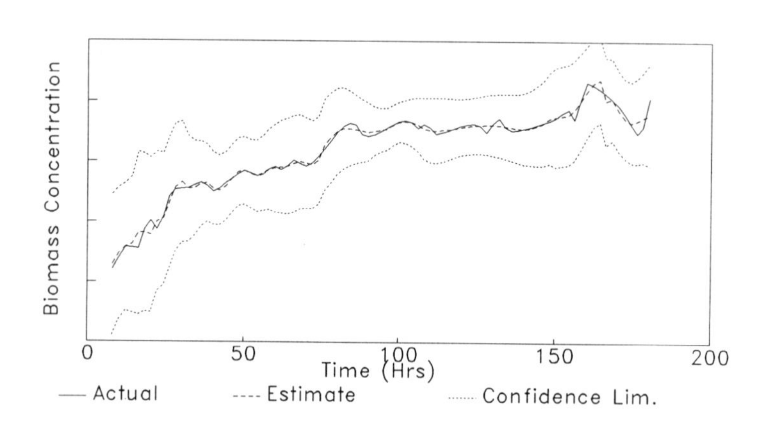

Figure 2. RBF: Training set #2

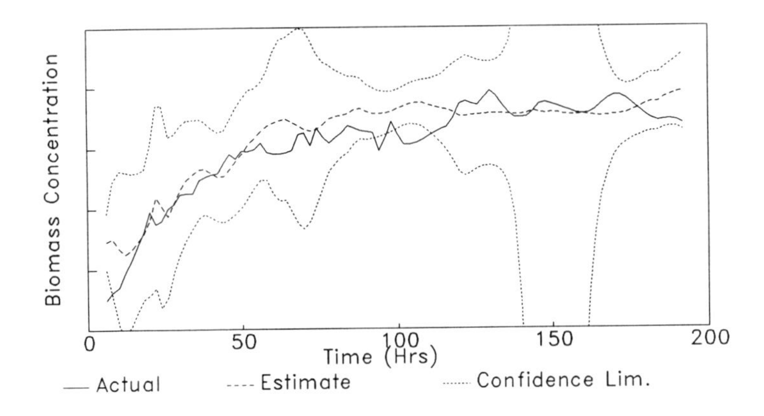

Figure 3. RBF: Propagation set

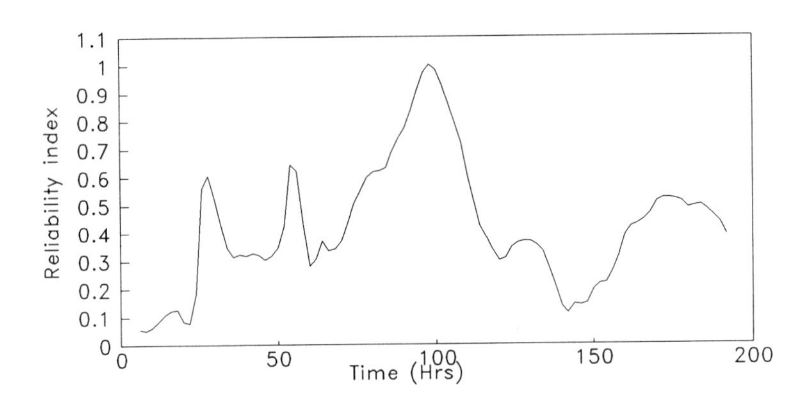

Figure 4. RBF: Reliability Index

output signal exactly, as is shown in figure 3. It picks up the trends very well and results are definitely good enough to be used for control purposes, especially since it is known that measurements tend to get more noisy towards the end. What is a remarkable feature in figure 3 is the way in which the Confidence Interval suddenly becomes very wide and then returns to the normal width again. Studying the training set can possibly provide an explanation for this phenomenon. At the point where the interval becomes wide, the two fermentations that make up the training set exhibit exactly opposing behaviour. Therefore the variance (derived from the predictive squared error) will be very large and the result is a very wide confidence interval. Also when looking at the reliability index in figure 4, it is clear that it reaches a local minimum where the confidence intervals are widest. The network indicates that is extrapolating up to a certain degree. If we then look at the result of the estimation, we find that the network is actually not doing bad at all. The training method, in order to minimise the squared error, has chosen a solution that is in the middle of the training set. The networks estimation of biomass at this point is in agreement with expert opinions of what is actually going on in the fermentation process.

b) The PLS Neural Network

Figures 5 to 7 show data from the penicillin process described earlier. The inputs time and CO2 are used to estimate biomass concentration. The first two process batches were used for training, the third for prediction.

These figures are obtained by using only the first of the possible PLS factors. To determine the number of PLS factor to be used for modelling purpose, it is instructive to examine the inner models created by the NNPLS algorithm for this data.

Figure 8 shows there is a clear relationship between the scores of the first factor and also shows the non-linear fit found for the first inner model. Clearly this factor should be included in the final model. The situation represented in figure 9 is rather different. Here the points are widely scattered and the NNPLS inner model does little to explain the relationship between these. In this particular case, the second inner model has been seen to add slightly more information which enables a better fit of the training data, but it does not contribute anything significant to the networks' ability to predict. That is why, as mentioned earlier, the second PLS factor has not been used. For other processes, adding extra factors actually degrades the performance capabilities. Note that this does not mean that one of the inputs can be neglected: each factor consists of a combination of all the inputs. More detailed analysis of the relationships between the inputs and the factors is needed to eliminate input variables.

The advantage of replacing the PLS linear, on nonlinear polynomial, inner model by a non-linear neural network accrues from the potential of being able to 'explain' complex nonlinear characteristics within a well-understood statistical framework. From the neural network stand-point, it provides a means of preprocessing the data so

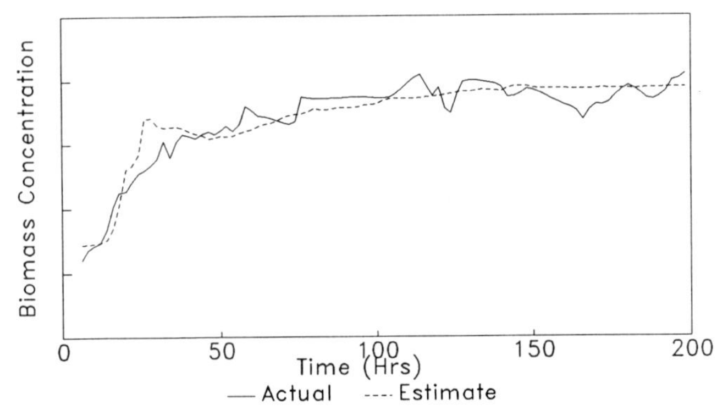

Figure 5. PLS–NN: Training set #1

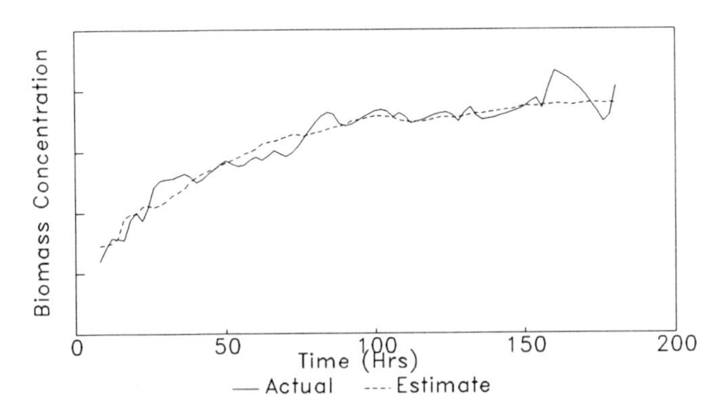

Figure 6. PLS–NN: Training set #2

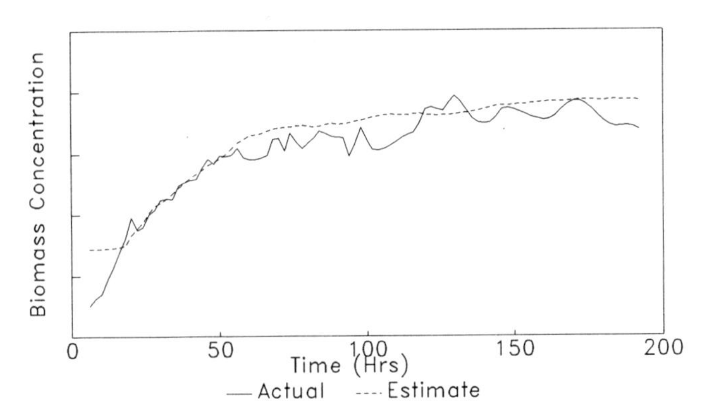

Figure 7. PLS–NN: Propagation Set

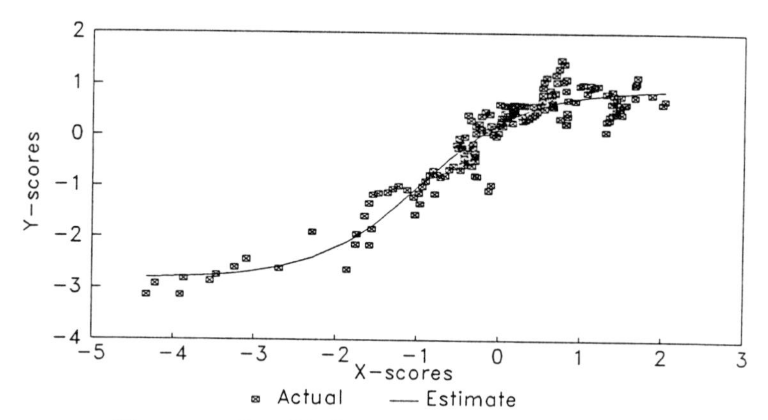

Figure 8. PLS−NN: First inner model

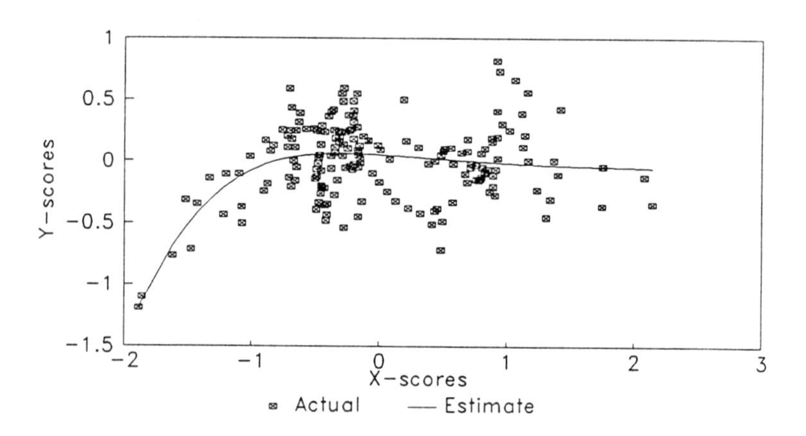

Figure 9. PLS−NN: Second inner model

that a series of small, simple neural networks can be used (each network has only one input and one output). Problems observed previously in normal feedforward networks where several of the inputs to a network are strongly related are overcome. Since factors are added sequentially, it is possible to examine the scores of each factor and decide whether or not it should be included in the final model.

c) The Feedforward Network

A feedforward artificial neural network (FANN) has been trained as shown in figure 10 and 11. A third (unseen) data set was then used to assess its performance (figure 12). It can be observed that in both training and propagation a smoothed estimation of biomass concentration is achieved. It is worth noting that dynamics and time delays can easily be introduce in the feedforward network, to provide for a better representation of the process. This is described in Di Massimo et al (1991).

A potential problem with using neural network model-base estimators is that network models might have been identified using data collected from the plant which may have some of its control loops still closed. The resulting model will then have been identified with correlated data and will not be representative of the underlying process behaviour. When such a model when used within a feedback control loop it will be subject to new process data that is further correlated and the model predictions will degrade. In this case it is important to identify a new network using the loop data now available and when this new network predictions are deemed better than those of the previous network model it should replace the 'old' model.

7.7 Static Networks versus Dynamic Networks

The basic feedforward network performs a non-linear transformation of input data in order to approximate output data. This results in a *static* network structure. In some situations such a steady state model may be appropriate, however, a significant number of projects require a description of system dynamics. One means of achieving dynamic description is to adopt a time series modelling approach and specify delayed as well as current inputs to the network. Alternatively, one can again consider parallels with the biological neuron to enhance the basic feedforward network. Terzuolo and Bayly (1968) were able to show that transfer of information along a synapse (the interconnections) had associated dynamics which could be modelled by a first order transfer function. Furthermore, Watanabe (1969) was able to demonstrate that a time delay was also associated with the transmission. Other workers have modelled the temporal properties of synapes and dentrites by Finite Impulse Response filters (FIR). Thus the model of the function of the individual neuron might be extended account for dynamics and dead-time. Whilst this might be biologically motivated, no claims are made as to its biological plausibility. However, from a engineering point of view it does acknowledge the importance of time delays in the modelling of dynamic processes.

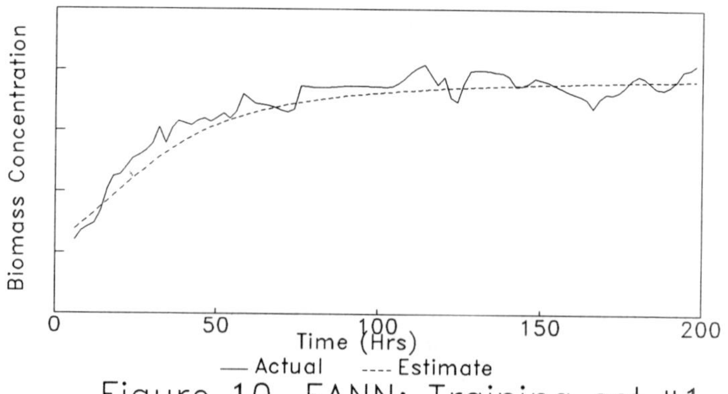

Figure 10. FANN: Training set #1

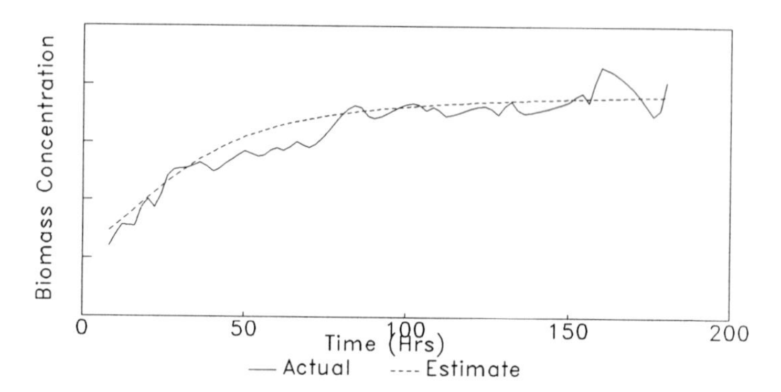

Figure 11. FANN: Training set #2

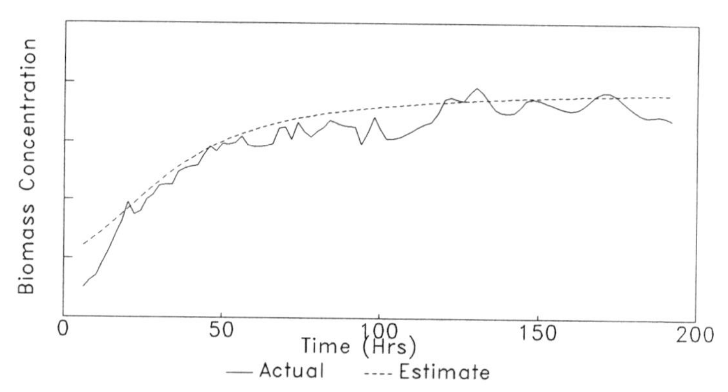

Figure 12. FANN: Propagation set

This modified neuron representation thus includes the necessary components to build representative non-linear process models. As with any approach to modelling, the model structure should be kept at its most simple, thus if dynamics or dead-time are not necessary to describe the data then their use within the network should be avoided. However, if the model requires the incorporation of such characteristics then there is justification and a means by which they may be introduced.

The use of network models as output, $y(t)$, predictors can be problematical if care is not taken in training the network properly. If the models are used to predict more than one-step-ahead in time, and have only been trained for such a task, then the ARMA time series and dynamic network approaches discussed above are not comparable. The one-step-ahead ARMA network model does not capture the process dynamics. Essentially, the autoregressive nature of the ARMA network results in the need to predict $y(t+n)$ from the estimates of $y(t+n-1)$. Errors in the estimate of y thus accumulate as the prediction horizon increases. Since the dynamic network model is not autoregressive, this problem does not arise and multi-step-ahead predictions are good. The problem of the ARMA network approach can, however, be overcome by minimising the network prediction error over time. That is the network training minimises $y(t)$ and all other output predictions up to a specified prediction horizon.

Recent studies have shown that the incorporation of dynamics into the network in this manner is highly beneficial in many real process application studies (eg. Montague et al, 1991, 1992).

7.8 Artificial Neural Networks in Control

If the neural'network model is of sufficient accuracy, then it should be possible to employ the model directly within a model based control strategy. A potentially useful algorithm may be one which minimises future output deviations from setpoint, whilst taking suitable account of the control sequence necessary to achieve this objective. this concept is not new and is common to most predictive control algorithms. However, the attraction of using the neural network instead of other model forms within the control strategy is the ability to effectively represent complex nonlinear systems.

7.9 Concluding Remarks

In this contribution, the concept of artificial neural networks was introduced and their suitability for solving some currently difficult process engineering modelling problems explored. One disadvantage of network models is that, in concert with their linear ARMAX and nonlinear NARMAX counterparts, inspection of the identified structure does not provide any mechanistic information about the system being modelled. The information held is distributed across all the network parameters and bears no relationship to the physical nature of the system. It is therefore difficult to interpret what the network has characterised and as a result predict its abilities and limitations in handling new unseen data. This is particularly important, especially since the problems escalate rapidly with large complex systems. In addition, the generation of training data sets capable of providing all the information necessary for the identification of large dimensioned networks, and the network training itself, could become prohibitive.

Conversely, mechanistic models can impart structure and meaning to the representation. Even though the difficulty in developing such models is often cited as a major reason for using a neural network approach, the use, association, or even incorporation of known process knowledge into a network structure is very appealing (Mavrovouniotis, 1991).

Applications to industrial process data reveals that given an appropriate topology the network could be trained to characterise the behaviour of all the systems considered. Artificial Neural Networks show good potential as 'software-sensors'. The additional benefits obtained from accurate output prediction, using secondary variables, allows for the effects of load disturbance compensation in a feedforward sense. Significant improvements in process regulation could be achieved when a network model is used directly for control purposes.

There is evidence that artificial neural networks can be a valuable tool in alleviating many current process engineering problems. However, it is stressed that the field is still very much in its infancy and many questions still have to be answered. Determining the 'best' network topology, if indeed there is one, is one example. What is best for one problem may well not be best for another. Currently, *ad hoc* procedures based on 'gut-feeling' tend to be used. This is not good and this arbitrary facet of an otherwise promising philosophy is a potential area of active research. A formalised technique for choosing the appropriate network topology is desirable. More analysis of network structure is needed to explore the relationship between the number of neurons required, to totally characterise the important process information, and the dimensionally finite topological space. There also no established methodology for determining the robustness and stability of neural networks in control loops. This is perhaps the one of the most important issues that has to be addressed before their full potential might be realised. Nevertheless, given the resources and effort that are currently being infused into both academic and commercial research in this area, it is anticipated that within the decade, neural networks will have been well established as a valuable tool.

7.10 Acknowledgements

The authors gratefully acknowledge the contributions from Christine Di Massimo, Anton Hofland and Aidan Saunders whose results form part of their Phd studies. The support of the Department of Chemical and Process Engineering, University of Newcastle and SmithKline Beecham is also gratefully acknowledged.

7.11 References

Akaike, H., (1974). "A New Look at the Statistical Model Identification", IEE Trans. Auto. Cont., AC-19, 6, pp 716-723.

Anderson, T.W., (1984), "Introduction to Multivariate Statistical Analysis", 2nd Ed., Wiley, New York.

Barron, A.R., (1984). "Predicted squared error: a criterion for automatic model selection", Self Organising Methods, S.J. Farlow (Ed.), pp 87-103.

Bhat N., Minderman, P. and McAvoy, T.J. (1989). "Use of neural nets for Modelling of Chemical Process Systems", Preprints IFAC Symp. Dycord+89, Maastricht, The Netherlands, Aug. 21-23, pp147-153

Bremermann, H.J. and Anderson, R.W. (1989). "An alternative to Back-Propagation: a simple rule for synaptic modification for neural net training and memory", Internal Report, Dept. of Maths, Uni. of California, Berkeley.

Chen, S., Billings, S.A., and Grant, P.M., (1990), "Nonlinear System Identification Using Neural networks", Int. J. Control, Vol 51, No. 6, pp 1191-1214.

Cybenco, G., (1989). "Approximations by Superpositions of a Sigmoidal Function", Math. Cont. Signal & Systems, 2, pp 303-314.

Di Massimo, C., Willis, M.J., Montague, G.A., Tham, M.T. and Morris, A.J. (1990). "On the applicability of neural networks in chemical process control", AIChE Annual Meeting, Chicago.

Di Massimo, C., Willis, M.J., Montague, G.A., Tham, M.T. and Morris, A.J. (1990). "Bioprocess model building using artificial neural networks", Bioprocess Engineering, 7, pp 77-82.

Hecht-Nielson R., (1989). "Theory of backpropagation neural networks". Proc Int. Conf. on Neural Networks, I, pp 593-611, IEEE Press, New York.

Holden A.V., (1976). "Models of the stochastic activity of neurones", Lecture Notes in Biomathematics, 12, Springer-Verlag.

Leonard, J. A. and Kramer, M.A. (1990) "Improvement of the back-propagation algorithm for training neural networks." Computers and Chem. Engng., 14, pp 337-341.

Leonard, J.A., and Kramer, M.A., and Ungar, L.H., (1991). "A Neural Network Architecture that Computes its Own Reliability", Comput. Chem. Engng.

Leontaritis, I.J., and Billings, S.A., (1985). "Input-output paramatric models for nonlinear systems, Part I: Deterministic nonlinear systems; Part II: Stochastic nonlinear systems, Int. Journal of Control, 41, pp 303-344.

Leontaritis, I.J., and Billings, S.A., (1987). "Model selection and validation methods for non-linear systems", Int. Journal of Control, 45, pp 311-341.

Ljung, L. and Soderstrom, T. (1983). "Theory and Practice of Recursive Identification", MIT Press.

Mavrovouniotis, M.L., (1991). "Hierarchical Neural Networks", Submitted for publication in Computers and Chemical Engineering.

Micchelli, C.A., (1986). "Interpolation of scattered data: distance matrices and conditionally positive definite functions", Constructive Approximation, 2, pp 11-22.

Montague G.A., Willis M.J., Morris A.J. and Tham M.T. (1991). "Artificial Neural Network based multivariable Predictive Control". ANN'91, Bournemouth, November

Montague G.A., Tham, M.T., Willis, M.J., and Morris, A.J., (1992), "Predictive Control of Distillation Columns Using Dynamic Neural Networks", 3rd IFAC Symposium DYCORD+'92, Maryland USA, April.

Narendra, K.S. and Parthasarathy, K., (1990), "Identification and Control of Dynamical systems using Neural Networks, IEEE Trans Neural Networks, 1, 1, pp4-27

Psaltis, D., Sideris, A. and Yamamura, A.A. (1988). "A multilayered neural network controller", IEEE Control Systems Magazine, April, pp17-21.

Powell, M.I.D., (1985). "Radial Basis Functions for Multivariable Interpolation: a review", IMA Conf. on Algorithms for the Approximation of Functions and Data, RMCS Shrivenham.

Powell, M.I.D., (1987). "Radial basis function approximations to polynomials", 12th Biennial Numerical Analysis Conf., Dundee, pp 223-241.

Qin, S-Z., and McAvoy, T.J., (1991). "Nonlinear PLS Modelling Using Neural Networks", Accepted for publication in Computers and Chemical Engineering.

Ricker, N.L., (1988), "The Use of Biased Least Squares Estimators for Parameters in Discrete Time Pulse-Response Models", Ind. Eng. Chem. Res., 27(2), p343.

Rumelhart, D.E. and McClelland, J.L. (1987). "Parallel Distributed Processing: Explorations in the Microstructure of Cognition", Vol.1, Chp. 8, MIT Press, Cambridge.

Terzuolo C.A. and Bayly E.J., (1968). "Data transmission between neurons", Kybernetik, 5, pp 83-84.

Tham. M.T., Morris, A.J., Montague, G.A., and Lant, P.A., (1991), "Soft Sensors for Process Estimation and Inferential Control", J Proc. Control, vol 1, pp 3-14.

Wang, Z., Tham, M.T., and Morris, A.J., (1991). "Multilayer Neural Networks: approximated canonical decompoisition of nonlinearity", Accepted for publication in International journal of Control, September 1991..

Wang, Z., Di Massimo, C., Tham, M.T., and Morris, A.J., (1992). "A Procedure for Determining the Topology of Feedforward Neural Networks", Submitted for publication to the IEEE Transactions.

Wantanabe, Y, (1969), ""Statistical measurement of signal Transmission in the central nervous system of the crayfish", Kybernetik, 6, pp 124-130.

Wise, B.M., and Ricker, N.L., (1990), "The Effect of Biased Regression on the Identification of FIR and ARX Models", AIChE Abstracts, 1990 Annual Meeting, Nov. Chicago USA.

Wold, S., (1987), "PLS Modelling with Latent Variables in Two or More Dimensions", Frankfurt PLS Meeting.

Wold, H., (1982), "Soft Modelling, The Basic Design and Some Extensions", Systems Under Indirect Observation, Joreskog and Wold (Eds.).

Chapter 8

Selection of neural network structures: some approximation theory guidelines

J. C. Mason and P. C. Parks

8.1 Introduction - artificial neural networks ("ANNs") in control

Control engineers have not been slow in making use of recent developments in artificial neural networks: a pioneering paper was written by Narendra and Parthasarathy (1990) and more recent developments are surveyed in the present proceedings.

Neural networks allow many of the ideas of system identification and of adaptive control originally applied to linear (or linearised) systems to be generalised, so as to cope with more severe nonlinearities. Such strong nonlinearities occur in a number of applications e.g. in robotics or in process control.

Two possible schemes for "direct" adaptive and "indirect" adaptive control are shown in Fig. 1 and other schemes will be found elsewhere in these proceedings, but in this contribution we shall concentrate on the modelling to be carried out by the artificial neural networks continued in the boxes C and IM in Fig. 1. For example the box IM in Fig. 1b seeks to build a non-linear dynamic model of the plant P in discrete-time as

$$y_t = f(y_{t-1}, y_{t-2}, \ldots y_{t-r}, u_{t-1}, u_{t-2}, \ldots u_{t-s})$$

where $\{y_t\}$ is the output and $\{u_t\}$ the input to the plant P which is to be modelled using the non-linear function f. An immediate problem which arises here is how large the integers r and s should be. They should be kept as small as possible to reduce the complexity of the neural network topology and the weight

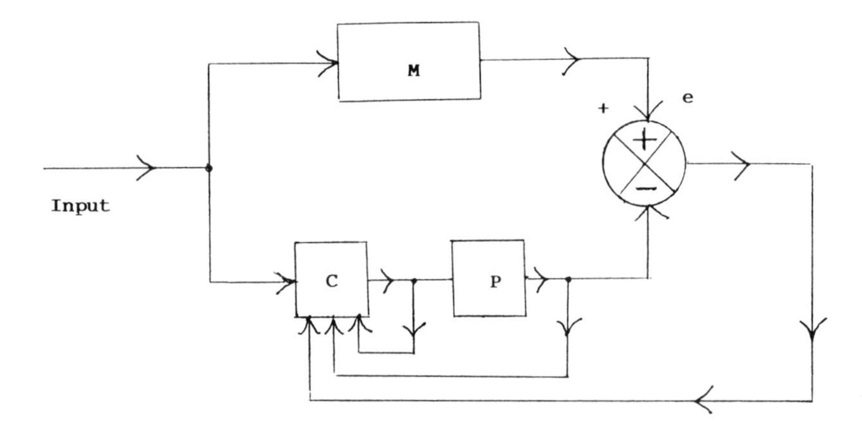

Fig. 1a "Direct" adaptive control

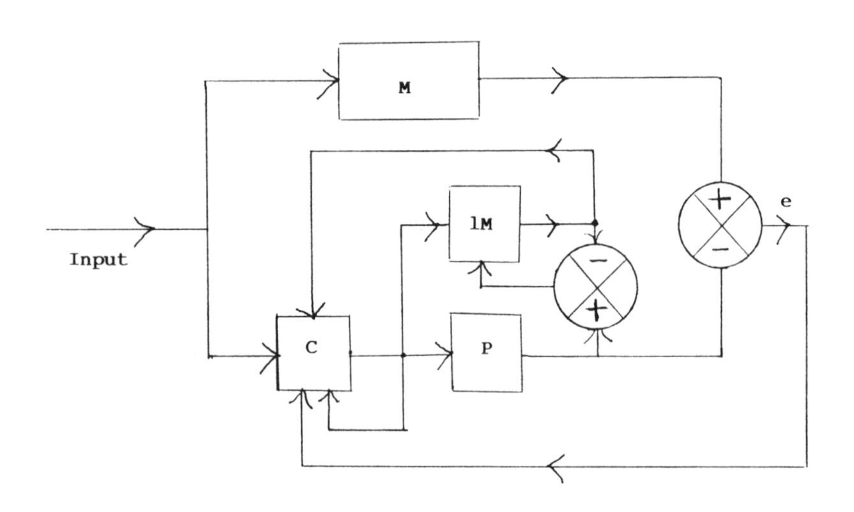

Fig. 1b "Indirect" adaptive control

M = Model, P = Plant, C = Controller

1M = identification model e = error signal

(ANNs replace C and 1M boxes in a neural network implementation)

learning procedure but large enough to model the significant dynamics of the plant P. Some theoretical and practical knowledge of P will be invaluable here; alternatively r and s may be chosen to be large initially and then subsequently "pruned" in a process later called "parsimony".

However, we shall now concentrate our thoughts on the general problem of modelling a single function of ℓ variables $f(x_1,x_2,...x_\ell)$.

8.2 Approximation of Functions

A neural network effectively creates a function of several variables from a number of sums and products of functions of one variable. It may use several layers of calculations, corresponding to the layers of the network, and it may modify the parameters in the resulting form of approximation by a learning or training procedure which passes through the network repeatedly or which moves forward and backward through the network.

Once the neural network has been designed and trained, it operates as a model function, outputting the value of a specific function F corresponding to the input of a value of each independent variable $x_1,x_2,...,x_\ell$. The function F so created does not coincide in general with the true function f being modelled, but acts as an approximation to it. Typically f is known at a discrete set of points $\underline{x}^{(1)},..., \underline{x}^{m}$, where

$$\underline{x}^{(i)} = (x_{1i}, x_{2i},...,x_{\ell i}), \quad (i=1,...,m)$$

corresponding to m sets of values of the vector variable

$$\underline{x} = (x_1, x_2,...x_\ell) ,$$

and so it is possible to compute the error f-F in the model F at each of the points $\underline{x}^{(i)}$.

The modelling process may thus be viewed in its total effect, as a problem in the "approximation of functions", and we may appeal to the mathematical principles of "approximation theory" to guide us in carrying out the process. For this purpose, there is a very substantial and rapidly growing literature available. The bulk of this literature is concerned with nonlinear functions, in the sense that f is not modelled as a hyperplane. Much of the literature is, however, restricted to "linear approximation" in the sense that the parameters, say

$$\underline{c} = (c_1, c_2, \ldots, c_n),$$

enter linearly into the form of approximation, though each parameter c_j multiplies a nonlinear basis function ϕ_j. Specifically

$$F(x) - \sum_{j-1}^{n} c_j \, \phi_j \, (x) \tag{2.1}$$

An "approximation problem" may then be posed, typically as one of the following three problems, based on the specification of a norm $\|.\|$ over the points $\underline{x}^{(i)}$.

(i) **Good Approximation Problem** Determine an approximation F, of the given form such that $\|f\text{-}F\|$ is acceptably small.
(ii) **Best Approximation Problem** Determine an F, F^{β} say, of the given form such that $\|f\text{-}F\|$ has its minimum value $\|f\text{-}F^{\beta}\|$.
(iii) **Near-Best Approximation Problem** Determine an F of the given form such that

$$\|f\text{-}F\| < (1+p). \, \|f\text{-}F^{\beta}\|$$

where F^{β} is a best approximation and p is acceptably small. (The norm is then within a relative distance p of its smallest value). (The maximum acceptable value for the constant p is up to the user. A value of 9 ensures a loss of accuracy of at most one significant figure, and a value of 0.1 ensures that the norm is minimised correct to about one significant figure.)

In practice, it is common to require that an approximation should be both good and best, or both good and near-best, in the senses of the above definitions. Obviously there is no great virtue in determining the best approximation of a form which is not capable of representing f accurately. Nor is it necessarily acceptable to obtain a good approximation, when an alternative procedure may yield a much better one or where a much smaller number of parameters may still provide a comparable one. The idea of a "near-best" approximation is introduced because it is not usually essential to find the actual "best" approximation, and indeed it may be time-consuming to do so. Although least squares methods readily yield best approximations in the ℓ_2 (or Euclidean) norm(see below), it is typically necessary to adopt iterative procedures to determine best approximations in other norms, and it may be sensible and acceptable to terminate these procedures in advance of their convergence.

The literature in approximation is vast and progressively more specialised. However, general discussions of approximation may be found in textbooks such as Powell (1981), and elementary introductions are given in many numerical analysis textbooks, such as in the last chapter of Mason (1984). State of the art discussions may be found in conference proceedings, such as the Texas "Approximation Theory" series - see Chui, Schumaker and Ward (1989) - and, on the more practical aspects, in the Cranfield "Algorithms for Approximation" conference series - see Mason and Cox (1990).

In posing a good, best, or near-best approximation problem as defined above, it is clear that a number of choices or specifications are needed, which make up what we call the "framework" of approximation. Without this framework, it is easy to attempt to solve the wrong problem. The framework consists of the approximation problem, together with three essential components: function, form, and norm.

(i) **Function** The given function $f(\underline{x})$ may be specified on a continuum or on a discrete data set $\{\underline{x}^{(1)},...,\underline{x}^{(m)}\}$. (In neural networks it is likely to be the latter). Moreover f may be exact or it may have errors in it, and in the latter case it is desirable to specify the nature and overall size of the data error. (For example, it might be assumed that the errors are normally distributed with a prescribed standard deviation).

In theoretical studies, the function f is often assumed to be from a vector space F of functions - which is infinite dimensional in the case of a continuum and of finite dimension m (and typically isomorphic to R^m) in the case of discrete data.

(ii) **Form of Approximation** The form is an approximation, such as (2.1), which includes a set of undetermined parameters c_j. It may be linear or nonlinear in the c_j, but in either case it can be identified with a vector $(c_1,...,c_n)$ in R^n. The form can be viewed as a vector space in the case of a linear form such as (2.1), with ϕ_j as the basis, and typically a subspace A of the function space F.

(iii) **Norm of Approximation** The norm is a measure of how well a specific approximation F of the given form matches the given function f. (We refer the reader to Powell (1981) for the definition of a norm). The most common general class of norms, in the case of discrete data, is the ℓ_p (Hölder) norm:

$$\| f-F \|_p - \left(\sum_{l-1}^{m} |f(x^{(l)}) - F(x^{(l)})|^p \right)^{1/p}$$

(2.2)

8.2.1 Choices of Norm

Assuming that f is given on a discrete data set $\underline{x}^{(1)},...,\underline{x}^{(m)}$, then the choice of the appropriate norm or measure of approximation is usually restricted to one of four, depending on the nature of the data and, to some extent, of the approximation algorithms available:

(*i*) $\ell_1 : \| f-F\|_{-1} - \sum_l | f(x^{(l)}) - F(x^{(l)}|$

(*ii*) $\ell_2 : \| f-F\|_2 = \left(\sum_i \left[f(x^{(i)}) - F(x^{(i)}) \right]^2 \right)^{1/2}$

(*iii*) $\ell_\infty : \| f-F\|_\infty = \max_i | f(x^{(i)}) - F(x^{(i)})|$ (*uniform norm*)

(*iv*) ***Smoothing (penalised least squares) measure***

$$\| f-F \|_2 + \lambda \int_{R^t} (PF)^2 dS$$

where P is a constraint operator, typically an appropriate second derivative operator.

Each of these choices has an ideal area of application. The ℓ_1 norm is excellent for data which have a small number of isolated wild points (or outliers), and tends to ignore such data. The ℓ_2 norm is best suited to data with errors in a normal (or statistically related) distribution, where the standard deviation in the error is not large and where approximations of at best comparable accuracy to the data are required. The ℓ_∞ norm is ideal for data which are exact, or very accurate, or have errors in a uniform distribution (such as data which have been rounded). Finally a smoothing measure is aimed at data with significant but statistically distributed noise, or at data with gaps (Daman and Mason (1987)). The inclusion of the integral term in the measure tends to "regularise" the data, by keeping certain derivatives small, and to fill gaps without undue oscillation.

In the context of neural networks, the choice of the ℓ_2 norm is almost universal, partly on the grounds that errors are present in the data, and partly because algorithms are seen to be simpler. However, excellent algorithms are available in libraries such as NAG for ℓ_∞ approximation, such as that of Barrodale and Phillips (1975), and for ℓ_1 approximation, such as that of Barrodale and Roberts (1974). Smoothing measures have been adopted, for example, by Poggio and Girosi (1989), and the reader should refer to discussions such as those of Von Golitschek and Schumaker (1990) and Daman and Mason (1987).

8.2.2 Choice of Form

There are a variety of popular choices of forms of approximation, all of which have been used in neural networks.

(i) **Algebraic Polynomial** (of degree n-1)

$$(a)\ F = \sum_{j=1}^{n} c_j\, x^{j-1}$$

$$(b)\ F = \sum_{j=1}^{n} c_j\, \phi_{j-1}(x)$$

where $\{\phi_{j-1}\}$ is an orthogonal polynomial system

(ii) **Trigonometric Polynomial** (of order n)

$$F = a_o + \sum_{j=1}^{n} (a_j \cos jx + b_j \sin jx) \quad (\textit{parameters } a_j, b_j)$$

(iii) **Spline of degree s** (knots $\xi_1, \ldots, \xi_r$)

There are 2 common forms used which, in the case of cubic splines (s=4), are as follows. (Splines of higher/lower degree may be defined similarly.)

$$(a)\ F = \sum_{j=1}^{4} c_j\, x^{j-1} + \sum_{j=5}^{r+4} c_j\, (x-\xi_{j-4})_+^s$$

(Truncated Power Form)

$$\text{where } (x-a)_+^3 = \begin{cases} 0 & \text{for } x < a \\ (x-a)^3 & \text{for } x > a \end{cases}$$

$$(b)\ F = \sum_{j-1}^{r+4} c_j\, B_j(x) \quad \text{(B-spline form)}$$

$$\text{where} B_j(x) - \sum_{k-o}^{4} d_k^{(j)} (x-\xi_{j+k-2})_+^3 \quad \text{(B-spline centred on } \xi_j)$$

$\xi_{-1}, \xi_0, \xi_{r+1}, \xi_{r+2}$ are added exterior or end point knots, and $d_k^{(j)}$ are so chosen that $B_j = 0$ for $x \geq \xi_{j+2}$. (The latter condition gives 4 equations for $d_k^{(j)}$. It remains to normalise B_j, and there are 2
standard ways of doing this - see de Boor (1978). (Note also that B_j vanishes already for $x \leq \xi_{j-2}$ and hence it has support only on $[\xi_{j-2}, \xi_{j+2}]$).

(iv) **Radial Basis Function** (Multivariate)

$$F - c_o + \sum_{j-1}^{n} c_j\, \phi(\|x - y^{(j)}\|) \tag{2.3}$$

where ϕ is a chosen function and $y^{(j)}$ are a set of chosen centres.

(v) **Ridge Function**

$$F - \sum_{j-1}^{n} c_j\, \phi_j(w_j^T x + \theta_j) \tag{2.4}$$

where $\underline{w}_j$, $\underline{x}$ are ℓ-vectors, θ_j is a scalar.

[Here, constant values of

$$z_j = w_j^T x + \theta_j = \sum_{l=1}^{\ell} w_{lj} x_l + \theta_j \tag{2.5}$$

represent hyperplanes on which F is constant, and hence a ridge is formed in F]

Typical choices of ϕ_j are **sigmoids** of the form $\phi_j = \sigma$, where

$$\text{either } \sigma(t) = \begin{cases} 1 & t>0 \\ 0 & t<0 \end{cases} \quad \text{or} \quad \sigma(t) = \frac{1}{1+e^{t}}$$

These sigmoids satisfy the boundary conditions: $\sigma(-\infty)=0$, $\sigma(+\infty)=1$.

The choices (iv) and (v) are those which appear to be most popular in current work on neural networks. The ridge function has been the traditional choice, but the radial basis function is becoming a popular competitor. A simple structure for the realisation of a **ridge function** has an input layer, a hidden layer, and an output layer. Here the variables $x_1, x_2, \ldots, x_\ell$ are inputs in the first layer, these are collected in the form $\underline{w}_j^T \underline{x} + \theta_j$ at the node j of the hidden layer, where they are processed by node function $\phi_j(.)$. They are then collected and the output at the 3rd layer is in the form (2.4). Suppose there are ℓ,n and p nodes in the 1st, 2nd and 3rd layers, respectively, then there are p output functions

$$F^{(k)} = \sum_{j=1}^{n} c_j^{(k)} \phi_j(z_j) \quad , (k=1,\ldots,p) \tag{2.6}$$

of the form (2.4)(2.5) with connection weights $c_j^{(k)}$ between 2nd and 3rd layers, and connection weights w_{ij} between 1st and 2nd layers. This structure is illustrated in Figure 2a. Strictly speaking, if $F^{(k)}$ approximates a known output $f^{(k)}$, say, then the

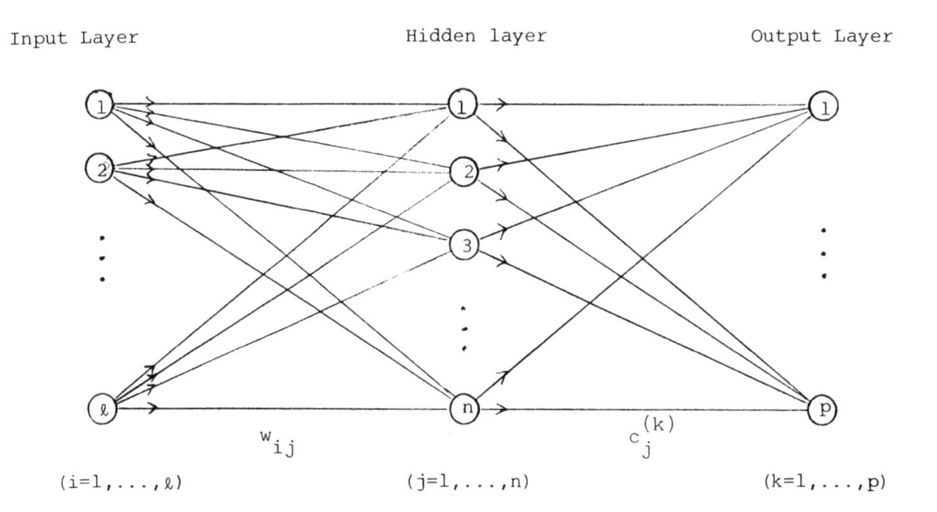

2(a) Hidden Layer - Ridge Functions

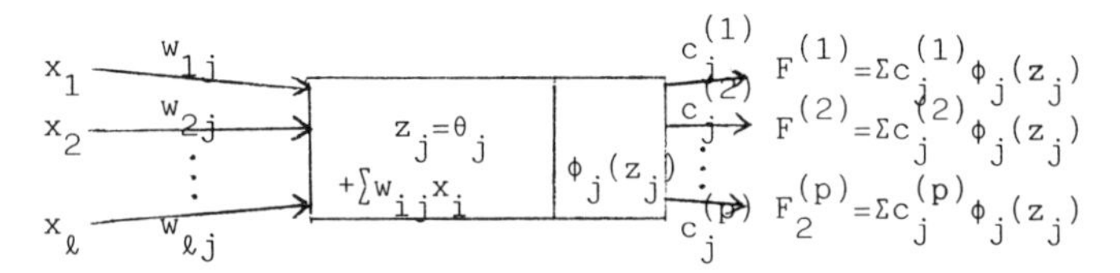

2(b) Hidden Layer - Radial Basis Functions

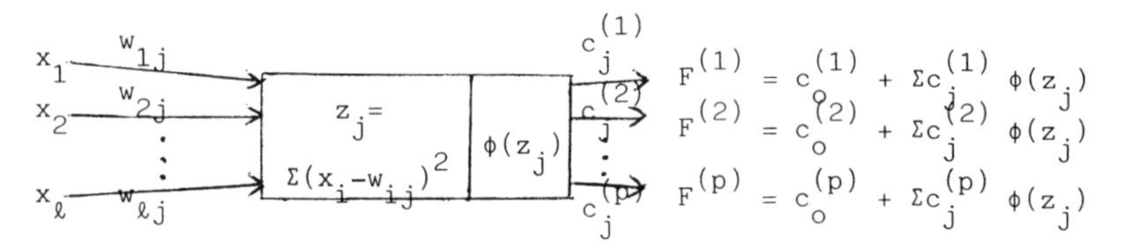

Figure 2 Feed Forward Networks

fitting problem consists of p simultaneous approximation problems $f^{(k)} \simeq F^{(k)}$ each involving a function of ℓ variables.

Further hidden layers could build more complicated functions than (2.4), for example one extra layer would form linear combinations of sigmoids whose arguments are formed from (2.4).

Radial basis functions have been recently adopted and may also be readily realised in a three layer structure, with input layer, one hidden layer, and output layer - as described by Broomhead and Lowe (1988). Again the input nodes contain the variables $x_1,...,x_\ell$. The hidden layer has n nodes, one for each centre $\underline{y}^j$, and each connection has a scalar y_{ij} assigned to it. Here y_{ij} is the ith component in $\underline{y}^j$, and its value links the ith input node to the jth hidden node. The fan-in to the jth hidden node is

$$z_j = \sum_{l=1}^{\ell} (x_l - w_{lj})^2 = \|x - y^{(j)}\|^2,$$

and the hidden layer applies the (radial) function ϕ to $(z_j)^{1/2}$. Thus the form (2.3) is the output at the third layer. Assuming again that there are p output nodes with output functions $F^{(k)}$ and weights $c_j^{(k)}$ (k=1,...,p) each of form (2.3). Then the connection weights in the network are $c_j^{(k)}$ between 2nd and 3rd layers and w_{ij} between 1st and 2nd layers. This structure is illustrated in Figure 2b. Radial basis functions have been adopted (possibly in modified form) by several neural network researchers in recent papers, such as Poggio and Girosi (1989), Chen, Billings, Cowan and Grant (1990), and Kramer (1991). Detailed discussions of radial basis functions, including choices of functions ϕ, are given by Powell (1987) and Buhmann and Powell (1990).

A number of radial functions were also proposed and tested in early practical discussions of Franke (1982). There are good arguments for using each of a number of choices of ϕ, and indeed all appear to perform with reasonable effectiveness and versatility. The choices that have good theoretical backing are as follows:

(a) $\phi(r) = r$
(b) $\phi(r) = r^3$
(c) $\phi(r) = r^2 \log r$ (thin plate splines)
(d) $\phi(r) = \exp(-r^2/2)$ (Gaussian)
(e) $\phi(r) = (r^2+c^2)^{½}$ (multiquadric)
(f) $\phi(r) = (r^2+c^2)^{-½}$ (inverse multiquadric)

For neural networks, thin plate splines have been adopted by Chen, Billings, Cowan and Grant (1990), and multiquadrics and Gaussians have been used by Broomhead and Lowe (1988).

Polynomials and spline functions, which have a long history in approximation theory have also been adopted for neural networks. Incidentally the forms (i) (ii) (iii), as we have defined them above, are given as functions of one variable x. However, analogous multivariate forms may be defined, typically by forming tensor products of univariate basis functions, such as

$$B_{J_1}(x_1)\ B_{J_2}\ (x_2) \ldots.\ B_{J_\ell}(x_\ell)$$

in the case of multivariate B-splines.

Orthogonal polynomials have two particular advantages. They provide explicit solutions to least squares problems (without the solution of simultaneous equations) in terms of ratios of inner products, and moreover they provide numerically well-conditioned families of basis functions. They are much more powerful in fitting multivariate functions than many people realise, and indeed high degree approximations may be successfully computed to fit relatively complicated functions. It is a great advantage to have data which lie on a mesh, or alternatively on lines, planes, or other structures. Bennell and Mason (1991) give a variety of powerful algorithms based on Chebyshev polynomials for bivariate approximation, which could certainly be extended to higher dimensions. Orthogonal polynomials have been adopted in a neural network context by Chen, Billings and Luo (1989). Incidentally, multivariate homogenous polynomials can be generated from a basis of ridge functions, namely

$$(\underline{w}^T\underline{x})^k \tag{2.7}$$

Results of Light (1991a) and Chui and Li (1991) establish that such basis functions span homogeneous polynomials and polynomials, respectively.

Most of the basis functions that we have discussed so far, namely (i) (ii) (iv) and (v) are defined globally. However, **B-splines**, which are the most popularly adopted basis functions for splines, have the potentially great advantage of being locally defined. Each B-spline has compact support and, for example, a cubic B-spline is only supported on 4 sub-intervals (connecting its knots) in each variable (see Fig. 3). This gives two computational advantages. First, B-spline algorithms lead to simultaneous equations with sparse matrices (typically banded matrices), and, second, their evaluation is very economical since it only adopts local information. A detailed discussion of splines and their properties is given in such texts as de Boor (1978) and Schumaker (1981). They have been adopted in a neural network context by Brown and Harris (1992), who use tensor product B-splines and claim to exploit their local properties.

The early 'CMAC' device of Albus (1975) can now be classified as belonging to an analogous "local" class.

8.2.3 Existence of good approximations

Possibly the most important problem in approximation theory, especially in the context of fitting high dimensional nonlinear functions which occur in neural networks, is that of establishing that good approximations exist. More precisely we need to show that the approximation space, for sufficiently high orders n, is dense (in a given norm) in the function space. Unless otherwise stated the maximum absolute error (uniform norm) is adopted as norm.

A fundamental result, which is also of great importance in many areas of analysis, concerns algebraic polynomials: **Weierstrass's 1st Theorem** states that algebraic polynomials are dense in the uniform norm in functions continuous on I_ℓ, the ℓ-dimensional cube $[0,1]^\ell$. A standard and constructive proof of this theorem is based on the representation of a continuous function in terms of **Bernstein polynomials.** An analogous result immediately follows for spline functions, when viewed as generalisations of polynomials.

In the context of neural networks, the main concern is centred on ridge functions and radial basis functions, and here significant results have been established recently.

Cybenko's Theorem [Cybenko (1989)]

Let σ be any continuous discriminatory sigmoidal function. Then ridge functions of the form (2.4) are dense in $C(I_\ell)$.

Here σ is said to be discriminatory if, for a measure μ(which is a signed regular Borel measure),

$$\int_{I_\ell} \sigma(w^T x+\theta)\ d\mu(x)=0$$

for all y in R^ℓ and θ in R implies that $\mu=0$.

This theorem has been strengthened by Chui and Li (1991), in the sense that, if the components of $\underline{w}$ and $\underline{\theta}$ are restricted to integer values, then the ridge functions are still dense in $C(I_\ell)$. Moreover, Chui and Li go beyond establishing existence, and indeed, for σ of bounded variation, they provide a construction of an arbitrarily accurate ridge function approximation (and hence a realization of a neural network with one hidden layer of arbitrary accuracy). Their construction is based on the use of a ridge polynomial, using the basis (2.4), and the expression of a continuous function in terms of Bernstein polynomials and hence ridge polynomials.

Mhaskar and Micchelli (1991) provide a more general result than that of Cybenko. They define **k^{th}-degree sigmoidal functions** with the property that

$$\lim_{x\to-\infty} x^{-k}\sigma(x) = 0 , \qquad \lim_{x\to\infty} x^{-k}\sigma(x) = 1$$

and which are bounded by a polynomial of degree at most k on R. They then show that such k^{th} degree sigmoidal functions are dense in continuous functions (on any compact set in R^{ℓ}).

Radial basis functions have also been studied in some depth. Recent treatments are given by Light (1991b), Dyn (1987), Dyn (1989), and Powell (1991). Light's discussion is detailed on the question of denseness. He shows in particular that the basis functions (a)(d)(e)(f) (i.e. all those listed above with the exception of $\phi(r)=r^3$ and thin plate splines), are each dense in continuous functions (on a compact set of R^{ℓ}). The latter two cases are discussed by Powell (1991), who shows that it is necessary to augment the basis by adding linear polynomials in order to ensure denseness.

Thus all the forms that we have listed in section 2.2 have the potential for approximating given multivariate functions (provided that in some cases some additional simple functions are added to the basis).

In concluding this section, it is probably appropriate to quote a famous result of Kolmorgorov (1957), that a continuous function of several variables may be represented by superpositions of continuous monotonic functions of one variable and the operation of addition. This result certainly supports the principle and potential of neural networks. However, it does not provide a realistically constructive realisation of a suitable network. The density results for specific forms, that we have quoted above, restrict the choice of network, adopt well understood forms of approximation and yet still guarantee the possibility of good approximations.

8.2.4 Approximation by interpolation

In the univariate case, the algebraic polynomials, trigonometric polynomials, and splines all have the property of providing unique interpolation on a set of distinct points equal in number to the number of approximation parameters. This is useful in itself in providing a potential algorithm, but it also implies that the basis forms a "Chebyshev set" (or "Haar system"), and hence that best approximations are unique in the ℓ_∞ norm.

However, in the multivariate case it is not usually possible to guarantee a unique interpolant for the above basis functions, and hence a best approximation is not necessarily unique.

Radial basis functions are a glowing exception in this area. Indeed it has now been established that all six forms (a) to (f) (augmented by linear polynomials in cases (b) (c)) uniquely interpolate any set of data on a distinct set of abscissae. All cases are covered in Light (1991b) and Powell (1991). The problem has a long history. Schoenberg (1946) showed indirectly that there was a unique interpolant in case (a) ($\phi = r$), but the first author to solve the problem in some generality appears to have been Micchelli (1986).

8.2.5 Existence and uniqueness of best approximation

Best approximations exist for all (linear) forms and norms that we have quoted above (the major requirement being that the search may be restricted to a compact set). Uniqueness also follows whenever the norm is "strict", and this covers the ℓ_p norm (2.2) for $1 < p < \infty$ and in particular the ℓ_2 norm.

We are thus left with some uncertainty in both ℓ_1 and ℓ_∞ norms. It is clear that best ℓ_1 approximations are not in general unique, and indeed there are many situations in which a continuum of parameter sets provide an infinite number of best approximations. Best ℓ_∞ approximations are unique if the basis functions form a Chebyshev set, and this is the case for **radial basis functions** (as a consequence of their interpolation property). However, there is no guarantee of uniqueness in ℓ_∞

for other forms.

In practice uniqueness is not necessarily a stumbling block in the implementation of algorithms and, for example, linear programming algorithms can be very effective for obtaining good ℓ_∞ approximations.

8.2.6 Parallel algorithms

Neural network approximation problems can involve very large numbers of parameters, possibly in the thousands or more. It is therefore highly desirable to consider adopting parallel processing procedures within the model and algorithm. Some useful progress has been made in this area, although, in our opinion, the subject is relatively poorly developed in the context of approximation algorithms. We shall therefore concentrate on drawing attention to relevant published work and potentially useful ideas, in the hope that this may lead to more mainstream neural networks approximation developments. Unless otherwise stated, all work is concerned with ℓ_2 norms.

A key parallel principle is that of **domain decomposition**, by which a domain is split up into many subdomains, on each of which a subproblem is solved in parallel. Such an approach, based at present on spline functions, has been studied by Montefusco and Guerrini (1991) (using L-splines and scattered data) and by Galligani, Ruggiero and Zama (1990) (using B-splines and meshes).

An important technique, which combines algorithm and architecture, is the use of a **systolic array** - a form of "data flow architecture" in which data are processed without storage as they are fed through an array of processing elements. Such an architecture has been adopted successfully for least squares approximation with radial basis functions by McWhirter, Broomhead and Shepherd (1991), and should in principle be readily extendable to other forms of approximation.

Locally supported basis functions would appear to be particularly suitable for the exploitation of parallel processing algorithms (such as domain decomposition), and indeed Brown and Harris (1992) claim to be exploiting parallelism in their B-spline model.

We should conclude this section by noting that there is, of course, massive "implicit parallelism" in a neural network. Great savings are immediately made if and when all operations at individual nodes in any layer are carried out in parallel.

8.2.7 Training/learning procedures and approximation algorithms

The fundamental design requirement in a neural network is to determine all the connection weights, such as w_{ij} and $c_j^{(k)}$ in Figures 2a and 2b, so that the generated output(s) match the required output(s) as closely as possible. This is an approximation problem, in that it is necessary to match $F(x_1,\ldots,x_\ell)$, over a set of data $\{(x_1,\ldots,x_\ell)\}$, by adopting an appropriate norm.

There are two distinct types of approximation which may be obtained, namely linear and nonlinear. If the weights w_{ij}, which link the first two layers, are regarded as fixed, and only the weights $c_j^{(k)}$ linking the second and third layers are allowed to vary, then the problem is a linear approximation problem. In that case, we may adopt a procedure such as the least squares method, and the problem then reduces to the least squares solution of an over-determined system of linear algebraic equations. (There is no shortage of good computer software packages for this problem - such as routine FO4 JGF in the NAG library, which obtains the ℓ_2 solution of minimal norm.) This is, for example, the type of approach proposed by Broomhead and Lowe (1988) for radial basis functions. Here the fixing of the weights w_{ij} corresponds to the fixing of the centres $y^{(j)}$, and so these need to be placed appropriately in advance.

However, if w_{ij} are also allowed to vary, then the problem becomes a nonlinear approximation problem, and it is inevitably much harder to solve. In the case of the ℓ_2 norm, it is a nonlinear least squares problem, which may be tackled by an optimization method designed for such problems (such as routine EO4 FDF in the NAG library). However, there is no guarantee that a global minimum of $\|f-F\|_2$ can be obtained by such a method in general, and many iterations may be needed even when convergence occurs.

The traditional approach in neural networks, which is in the end equivalent to solving the nonlinear approximation problem, is to select the weights w_{ij} and $c_i^{(k)}$ by an iterative "training" procedure, starting from some initial choice, which may even by random. A feedback approach for achieving this is discussed, for example, by Chen, Billings, Cowan and Grant (1990), and is based on a steepest descent or related procedure for solving the underlying optimisation problem. Again many iterations may be needed and there is no absolute guarantee of success.

8.2.7.1 Linear or nonlinear approximation?

Which of the two procedures should we adopt? Should we fix w_{ij} and solve a linear problem, or solve the full nonlinear problem by leaving w_{ij} free? This is a matter of opinion, depending to a great extent on the experience and/or knowledge that the user has in fixing the relevant parameters and on the ability of the resulting linear form to approximate the output well.

There is not as yet a great wealth of experience in the solution of the full nonlinear approximation problem in the cases of ridge functions (and other multi layer perceptrons) or radial basis functions, and indeed much of the existing experience lies with neural networks researchers. It is open to question whether or not current learning algorithms can be used with any confidence. However, there is considerable experience amongst approximation specialists in the solution of the nonlinear problem of fitting a multivariate **spline function**, using a basis of tensor product B-splines (2.6). Indeed, even in the one dimensional case, the problem of optimising the approximation with respect to free choices of both knots ξ_i and coefficients c_j has been found to be an extremely difficult task from a numerical point of view. It has been found, in particular, that the determination of an optimal set of knots ξ_j is typically a highly ill-conditioned problem. A great deal of time has been wasted by researchers over the past 20 years or so in futile attempts to obtain good "free knot" algorithms.

For this reason, the preferred approach to spline approximation is to fix knots by a selective or adaptive procedure, at each step of which a linear approximation problem is solved for the coefficients c_j. The problem of selecting good knots is not an easy problem either, but some progress has been made and, for example, knot insertion and deletion algorithms are offered by Cox, Harris and Jones (1990) and by Lyche and Morken (1990).

Some success has recently been claimed by Loach (1990) in solving the nonlinear approximation problem, in which splines are replaced by continuous piecewise polynomials (with less restrictive derivative continuity requirements). They adopt a dynamic programming procedure for solving the relevant nonlinear optimisation problem.

The lessons to be learned from splines are clear. Before embarking too enthusiastically on the nonlinear approximation problem, it is essential to study the conditioning of this problem. We are not aware that such studies have been undertaken as a routine matter in neural networks. In any case, there is much to be said in favour of the alternative use of selection procedures for determining good "nonlinear parameter" w_i, such as radial basis centres, so that the resulting approximation problem may become linear. However, there is a dearth of such selection procedures at present for both radial basis functions and ridge functions, and we believe that this is an area that deserves far more attention. It is not at all obvious, for example, where to place radial basis centres in problems of high dimension.

8.2.7.2 Economisation of parameters - parsimony

In addition to being highly nonlinear, the approximation

problem (in many dimensions) is also a very large one, and it is not impossible to be faced with thousands of connection weights. Indeed, since massively parallel computation is envisaged, there is a temptation to adopt a significant number of nodes in each layer of the network. We are then faced with a very challenging task - the solution of an optimization problem with many parameters - and our chances of success in a reasonable timescale are greatly reduced.

The procedure generally adopted by researchers, such as Chen et al (1990) has been to aim for "parsimony", by progressively eliminating those parameters that appear to be making insignificant contributions to the approximation. Indeed Chen et al have provided their own rigorous procedures fo deciding when to eliminate certain parameters in the case of radical basis function expansions. This is certainly an intelligent approach, and probably an essential one, but there are also potential snags in the approach. For example, a parameter may make a small contribution at one stage of an (interative) approximation procedure and then make a much larger contribution at a later stage. However, this is an area which also deserves considerable attention. Indeed, surprisingly little practical attention has been given by approximation specialists to the notion of parsimony, even for one-dimensional problems. There has, however, been some significant theoretical work on the use of "incomplete polynomials", in which certain terms are missing, by workers such as Saff (1983), and so there is some theoretical backing to such an approach.

8.2.8 Practicalities of approximation

In concluding this discussion of approximation, it is appropriate to point out various practical requirements which correspond to the various choices of forms of approximation, and which may influence the suitability of these choices. This leads us to attempt to compare the merits of different forms for use in neural networks.

Spline functions require a mesh of knots to be prescribed, and they effectively involve the subdivision of the approximation domain by some multivariate form of rectangulation. In a many-variable problem this can involve an enormous number of subdomains, and so the problem is potentially very cumbersome. On the other hand, B-splines have locally compact support, and so savings may be made by restricting computation to local areas of the network. Moreover parallel algorithms may well be based on the inherent domain decomposition. (These points have essentially been noted by Brown and Harris (1992).

Ridge functions are interesting in that they effectively create a function of many variables from functions of one variable. They have a natural and historical role in neural networks and are relatively convenient to adopt. They are less well known in mainstream approximation theory than other forms and, from that viewpoint might be regarded as somewhat unorthodox. They have global, rather than local support.

Radial basis functions have some unique features that stand out in fitting multivariate functions. They can uniquely interpolate, they posess unique best ℓ_∞ approximations, and they do not require the division of the domain into a mesh of subdomains. What they do, however, depend upon is an appropriate choice of n centres $\underline{y}_i$ (where n is the number of parameters), and this is both their greatest strength and greatest weakness. If centres are well chosen, then a small number of parameters may be required. Indeed, Chen et al (1990) claim that their algorithm can reduce the number of centres to be selected by an "orthogonal-forward-regression". On the other hand, it is hard to envisage a natural choice of centres; if a mesh of centres is chosen, then any advantage in efficiency over other forms may be lost. Some of the data abscissae might be chosen as centres, but there is no requirement that this should be the case. Would it be advantageous to adopt a technique analogous to those used in multivariate quadrature of high dimension, where the function is sampled by a Monte-Carlo or random-walk procedure in multivariate space?

Our instinct is that radial basis functions may just have the edge at present, since they are viewed with some favour by both neural network and approximation specialists. Ridge functions are popular in neural networks and may perhaps be the most natural choice, while spline functions are strongly supported in approximation theory for their local properties and versatility. These two latter forms are therefore strong contenders also.

Two other aspects which need to be taken into account are parallel processing and network training. The extent to which these aspects may be effeciently taken into account may well hold the balance in choosing finally between these three leading contenders.

Looking to the future, we should like to point out some possible advantages in using "local" functions such as B-splines rather than "global" functions such as sigmoids. "Local" functions are known more technically as functions with "compact support" which means that they are zero except in a limited finite interval in their argument (or arguments). Some one-dimensional examples are drawn in Fig. 3. These examples include rectangular functions (i.e. degree zero B-splines), triangular functions (i.e. linear B-splines), and cubic B-splines. They may be used with more overlapping than in Fig. 3 and may be generalised to 2 or more dimensions. Overlapping rectangular function in ℓ dimensions is used in the "Cerebellar Model Articulation Controller (CMAC)" developed by J.S. Albus (1975) in the early 1970's. (For a more recent exposition of his idea see Mischo et al (1991)).

Not only does this idea have a sound basis in the theory of spline approximation (see de Boor (1978)) but in the learning process to adjust the coefficients of the linear combination of these functions, from which the estimate F of f is formed, only a finite number of weights is adjusted at each stage of this learning process. Moreover a particularly simple projection algorithm, originally due to S. Kaczmarz (1937), is used to adjust those weights. The simplicity and speed of this algorithm lends itself to real-time on-line implementation, of the greatest importance for adaptive control using neural network techniques.

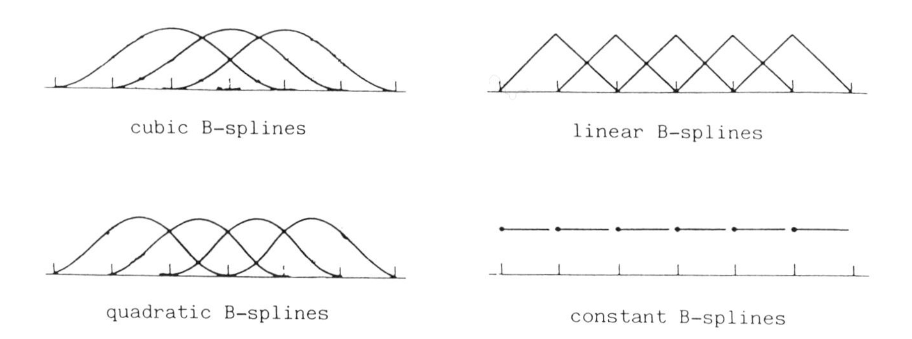

Figure 3 Some Local Basis Functions

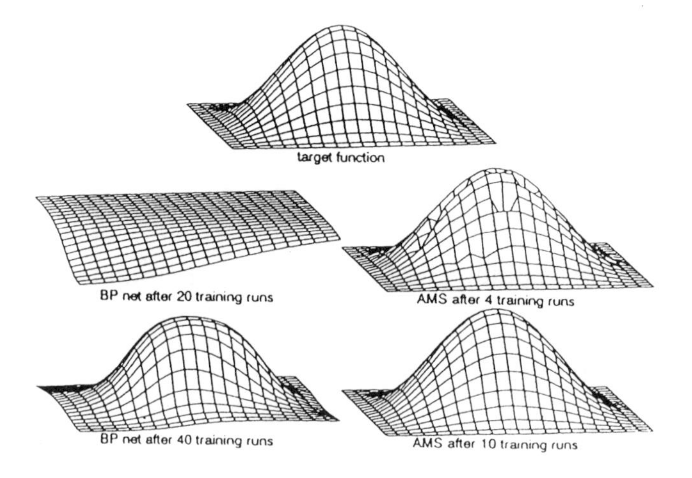

	AMS	Backpropagation Network
Test function	$\sin^2(x_1/\pi)\cdot\sin^2(x_2/\pi),\ x_{1,2}\in[0,1]$	
Resolution	1000×1000 (10^6 possible input points)	
No. of points per training run	1000 (randomly distributed)	
Parameters	generalization $\varrho = 32$	$2n+1=5$ hidden neurons

Fig.4 A Comparison of convergence of AMS and backpropagation networks

(32 overlapping local functions versus 5 sigmoids)

8.3 Convergence of weight training algorithms

The various schemes outlined in section 8.2 above can affect dramatically the convergence of the weight training algorithms. There is convincing evidence that modelling with "local functions" as opposed to "global functions" results in faster convergence of the weight training process. Fig. 4, taken from Mischo, Hormel and Tolle (1991), compares convergence of "ARMS" (Associative Memory Systems") modelling using 32 overlapping local functions with a (2-5-1) neural network (i.e. 2 input nodes, 5 hidden nodes and 1 output node) using the well-known back-propagation weight training algorithm (McClelland, Rumelhart et al 1986). Here a factor of x1/10 in convergence time is claimed in favour of "local" functions. Moody and Darken (1988) claim even higher factors e.g. x1/1000. This convergence speed could be a crucial deciding factor in on-line real-time closed loop use of ANNs in adaptive control loops, since we know that for more elementary linear control loops containing time delays that long delays usually lead to unstable closed loops.

8.4 Conclusions

What does approximation theory have to offer workers in neural networks?

1) Approximation theory offers the "alternative" approach of adopting a general purpose optimization method to solve the relevant nonlinear approximation problem, in addition to the usual feed-forward and backward propogation procedures.

2) Locally supported basis functions, such as B-splines, deserve to be given more attention. They have the potential for giving considerable savings, and they are also known to be versatile in fitting functions which vary in their behaviour locally over the domain.

3) For high-dimensional problems, radial basis functions are potentially valuable, since they may be based on a limited number of centres, which do not have to be placed on a grid

throughout the domain (as splines generally do). However, this advantage is only a real one if a good selection procedure is available for centres.

4) Researchers in approximation theory and algorithms have significant contributions to make in neural networks. Existence of good/best approximations has been fundamental, but attention now needs to be directed towards (a) the development of algorithms, (b) the rate of convergence of the approximation (form) as the number of parameters increase, and (c) the integration of approximation methodology into feed-forward, back propogation and related procedures.

REFERENCES

[1] J S Albus "A new approach to manipulator control: the cerebellar model articulation controller (CMAC)" ASME Transactions Series G, J of Dynamic Systems, Measurement and Control, 97 220-227, 1975.

[2] I Barrodale and C Phillips, Solution of an over-determined system of linear equations in the Chebyshev norm. ACM TOMS 1 (1975), 264-270.

[3] I Barrodale and F D K Roberts, Algorithm 478: solution of an overdetermined system of linear equations in the ℓ_1 norm. Comm. ACM 17 (1974), 319-320.

[4] R P Bennell and J C Mason, Bivariate orthogonal polynomial approximation to curves of data. In: "Orthogonal Polynomials and their applications", C Brezinski, L Gari and A Ronveaux (Eds), J C Balzer Pub. Co., IMACS (1991), 177-183.

[5] C de Boor, "A Practical Guide to Splines", Springer Verlag, New York, 1978.

[6] M Brown and C J Harris, The B-spline neurocontroller. In: "Parallel Processing in Control", E Rogers (Ed), Prentice Hall, 1992.

[7] D S Broomhead and D Lowe, Multivariable functional interpolation and adaptive networks. Complex Systems 2 (1988), 321-355.

[8] M D Buhmann and M J D Powell, Radial basis function interpolation on an infinite regular grid. In: "Algorithms for Approximation 2", J C Mason and M G Cox (Eds.),

Chapman and Hall, London, 1990, pp 146-169.
[9] S Chen, S A Billings, C F N Cowan and P M Grant, Practical identification of NARMAX models using radial basis functions. Int. J. Control 52 (1990), 1327-1350.
[10] S Chen, C A Billings and W Luo, Orthogonal least squares methods and their application to nonlinear system identification. Int. J. of Control 50 (1989), 1873-1896.
[11] C K Chui and X Li, Realization of neural networks with one hidden layer. Report 244, Center for Approximation Theory, Texas A and M University, March 1991.
[12] C K Chui, L L Schumaker and J D Ward (Eds), "Approximation Theory 6", Academic Press, New York, 1989.
[13] M G Cox, P M Harris and H M Jones. A knot placement strategy for least squares spline fitting erased on the use of local polynomial approximations. In: "Algorithms for Approximation 2", J C Mason and M G Cox (Eds), Chapman and Hall (1990) 37-45.
[14] G Cybenko, Approximation by superpositions of a sigmoidal function. Math. Control Signals Systems 2 (1989), 303-314.
[15] A E Daman and J C Mason, A generalised cross-validation method for meteorological data with gaps. in: "Algorithms for Approximation", J C Mason and M G Cox (Eds), Clarendon Press, Oxford, 1987 pp595-610.
[16] N Dyn, Interpolation of scattered data by radial functions. In: "Topics in Multivariate Approximation", C K Chui, L L Schumaker and F Utreras (Eds), Academic Press, New York, 1987, pp 47-61.
[17] N Dyn, Interpolation and approximation by radial and related functions. In: "Approximation Theory 6", C K Chui, L L Schumaker and J D Ward (Eds), Academic Press, New York, 1989, pp211-234.
[18] R Franke, Scattered data interpolation: tests of some methods. Math. and Computing 38 (1982), 181-200.
[19] I Galligani, V Ruggiero and F Zama, Solutions of the equality-constrained image restoration problems on a vector computer - In: "Parallel Processing 89", D J Evans, G R Joubert and F J Peters (Eds), Elsevier Pub. Co., 1990.

[20] S Kaczmarz "Angenäherte Auflösung von Systemers Linearer Glaichungen", Bull. Int. de l'Academie Polonaise des Sciences et des Lettres, Cl. d. Sc. Mathém. A, 355-357, Cracovie, 1937.

[21] A N Kolmogorov, On the representation of continuous functions of many variables by superposition of continuous functions of one variable and addition. Dokl. Akad. Nauk SSSR 114 (1957), 953-956.

[22] M A Kramer, Data analysis and system modelling with autoassociative and validity index networks. Proc. Int. Symposium "Neural Networks and Engineering Applications", Newcastle, Oct. 1991.

[23] W A Light, Private Communication, 1991(a)

[24] W A Light, Some aspects of radial basis function approximation (In press), 1991 (b).

[25] P D Loach, Best least squares approximation using continuous piecewise polynomials with free knots, PhD Thesis, Bristol University, 1990.

[26] T Lyche and K Morken, A discrete approach to knot removal and degree reduction algorithms for splines. In: "Algorithms for Approximation", J C Mason and M G Cox (Eds), Clarendon Press, Oxford (1987), 67-82.

[27] J C Mason, "BASIC Matrix Methods", Butterworths, 1984.

[28] J C Mason and M G Cox (Eds), "Algorithms for Approximation 2", Chapman and Hall, London, 1990.

[29] J L McClelland, D E Rumelhart (et al), "Parallel distributed processing", 2 vols. M.I.T. Press, Cambridge, Mass., 1986.

[30] J G McWhirter, D S Broomhead and T J Shepherd, A systolic array for nonlinear adaptive filtering and pattern recognition. J. of VLSI Signal Processing 3 (1991), 69-75.

[31] H N Mhaskar and C A Micchelli, Approximation by superposition of a sigmoidal function. (In press), 1991.

[32] C A Micchelli, Interpolation of scattered data: distance matrices and conditionally positive definite functions. Constructive Approx. 2 (1986) 11-22.

[33] W S Mischo, M Hormel, H Tolle, "Neurally inspired associative memories for learning control: a comparison" in "Artificial Neural Networks" (ed T Kohonen et al) Vol. 2, 1241-1244, Elsevier Science Publishers (North Holland), 1991.

[34] L B Montefusco and C Guerrini, Domain decomposition methods for scattered data approximation on a distributed memory multiprocessor. Parallel Computing 17 (1991), 253-263.

[35] J Moody, C Darken "Learning with localised receptive fields", Proc. 1988 Connectionist Models Summer School, 1-11, Morgan Kaufman Publishers, San Mafeo, California, 1988.

[36] K S Narendra, K Parthasarathy, "Identification and control of dynamical systems using neural networks" IEEE trans. on Neural Networks, 1, 4-27, 1990.

[36] T Poggio and F Girosi, A theory of networks for approximation and learning. AI Memo No 1140, MIT AI Laboratory, July 1989.

[37] M J D Powell, "Approximation Theory and Methods, Cambridge University Press, 1981.

[38] M J D Powell, Radial basis functions for multivariable interpolation In: "Algorithms for Approximation", J C Mason and M G Cox (Eds), Clarendon Press, Oxford, 1987, pp 143-168.

[39] M J D Powell, Radial basis functions in 1990. In: "Advances in Numerical Analysis Vol. II", Oxford University Press, 1991, pp105-210.

[40] E B Saff, Incomplete and orthogonal polynomials. In: Approximation Theory 4, C K Chui, L L Schumaker and J D Ward (Eds), Academic Press (1983), pp219-256.

[41] I J Schoenberg, Contributions to the problem of approximation of equidistant data by analytic functions. A B Quarterly Appl. Math. 4 (1946), 45-99 and 112-141.

[42] L L Schumaker, "Spline Functions: Basic Theory", Wiley, 1981.

[43] M Von Golitschek and L L Schumaker, Data fitting by penalized least squares. In: "Algorithms for Approximation 2", J C Mason and M G Cox (Eds), Chapman and Hall, London, 1990, pp210-227.

Chapter 9

Neural networks and system identification

S. Billings and S. Chen

9.1 Introduction

Neural networks have become a very fashionable area of research with a range of potential applications that spans AI, engineering and science. All the applications are dependent upon training the network with illustrative examples and this involves adjusting the weights which define the strength of connection between the neurons in the network. This can often be interpreted as a system identification problem with the advantage that many of the ideas and results from estimation theory can be applied to provide insight into the neural network problem irrespective of the specific application.

Feed forward neural networks, where the input feeds forward through the layers to the output, have been applied to system identification and signal processing problems by several authors (eg. Lapedes and Farber 1988, Narendra and Parthasarathy 1990,Brown and Harris (1992), Billings et al 1991 a, b and Chen et al 1990 a-d,1991, 1992) and the present study continues this theme. Three network architectures the multilayered perceptron, the radial basis function network and the functional link network are discussed and new training algorithms are introduced. A recursive prediction error algorithm is described as an alternative to back propagation for the multilayered perceptron. Two new learning algorithms for radial basis function networks, which incorporate procedures for selecting the basis function centres, are discussed and the extension of these ideas to the functional link network is described.

Feed forward neural networks are of course just another functional expansion which relates input variables to output variables and this interpretation means that most of the results from system identification(Ljung and Soderstrom 1983) can usefully be applied to measure, interpret and improve network performance. These concepts and ideas are discussed in an attempt to at least partially answer questions such as:- how to assign input nodes, does network performance improve with increasing network complexity, is model validation useful, will noisy measurements affect network performance, is it possible to detect poor network performance, can we judge when the generalisation properties of a network will be good or bad and so on.

While almost all of the results are applicable for alternative learning algorithms and should apply to the range of problems for which neural networks have been considered a discussion that relates specifically to the identification of nonlinear systems is also included.

Throughout the algorithms are compared and illustrated using examples based on data from both simulated and real systems.

9.2 Problem Formulation

Consider the nonlinear relationship

$$y(t) = f(y(t-1),\ldots,y(t-n_y),u(t-1),\ldots,u(t-n_u))+e(t) \qquad (9.2.1)$$

where

$$y(t) = [y_1(t)\ldots y_m(t)]^T,\ u(t) = [u_1(t)\ldots u_r(t)]^T,\ e(t) = [e_1(t)\ldots e_m(t)]^T \qquad (9.2.2)$$

are the system output, input and noise vectors respectively and f(.) is some vector valued nonlinear function. In the realistic case where the output is corrupted by noise lagged noise terms have to be included within f(.) and this defines the NARMAX model (<u>N</u>onlinear <u>A</u>uto<u>R</u>egressive <u>M</u>oving <u>A</u>verage model with e<u>X</u>ogenous inputs) introduced by Billings and Leontaritis (1981) ,Leontaritis and Billings (1985) and studied extensively in nonlinear system identification (Chen and Billings 1989). In an attempt to keep the concepts as simple as possible most of the current analysis will relate to the model of eqn (9.2.1) with the aim of approximating the underlying dynamics f(.) using neural networks. Throughout the network input vector will be defined as

$$\mathbf{x}(t) = [y^T(t-1)\ldots y^T(t-n_y)\ u^T(t-1)\ldots u^T(t-n_u)]^T \qquad (9.2.3)$$

with dimension $n_I = m x n_y + r x n_u$, where the one step ahead predicted output is expressed as

$$\hat{y}(t) = \hat{f}(\mathbf{x}(t)) \qquad (9.2.4)$$

and the model predicted output as

$$\hat{y}_d(t) = \hat{f}(\hat{y}_d(t-1), \ldots, \hat{y}_d(t-n_y), u(t-1), \ldots, u(t-n_u)) \qquad (9.2.5)$$

9.3 Learning Algorithms for Multilayered Neural Networks

9.3.1 The Multilayered Perceptron

A neural network is a massively parallel interconnected network of elementary units called neurons. The inputs to each neuron are combined and the neuron produces an output if the sum of inputs exceeds a threshold value. A feed forward neural network is made up of layers of neurons between the input and output layers called hidden layers with connections between neurons of intermediate layers. The general topology of a multilayered neural network is illustrated in Figure 9.1.

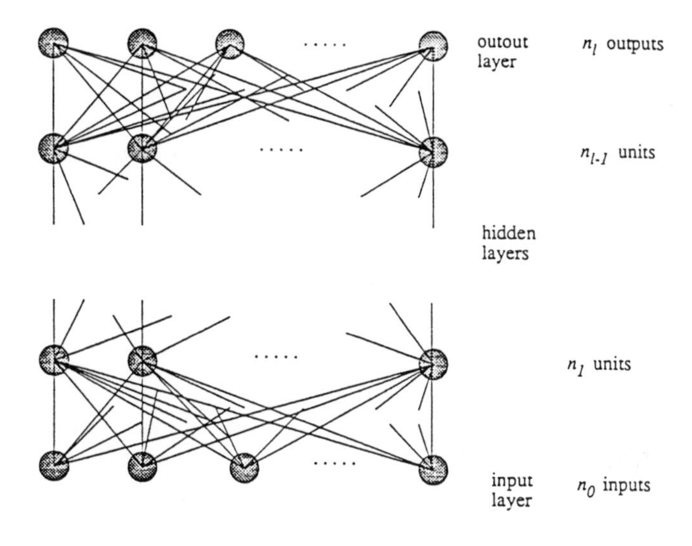

Figure 9.1 A Multilayered Neural Network

The input layer acts as an input data holder which distributes the inputs to the first layer. The outputs from the first layer nodes then become inputs to the second layer and so on. The last layer acts as the network output layer and all the other layers below it are called hidden layers. A basic element of the network, the i'th neutron in the k'th layer, illustrated in Figure 9.2, incorporates the combining and activation functions.

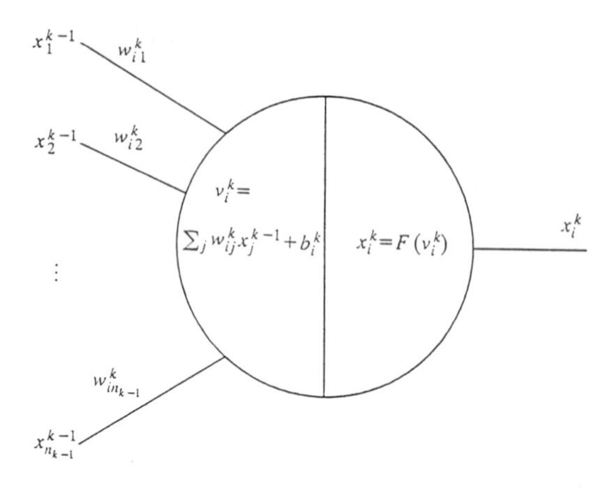

Figure 2 A Hidden Neuron i of Layer k

The combining function produces an activation for the neuron

$$v_i^k(t) = \sum_{j=1}^{n_{k-1}} w_{ij}^k x_j^{k-1}(t) + b_i^k \tag{9.3.1}$$

$$x_i^k(t) = F(v_i^k(t)) \tag{9.3.2}$$

where w_{ij}^k and b^k are the connection weights and threshold and F(.) is the activation function.

In applications to modelling nonlinear dynamic systems of the form of eqn (9.2.1), the network input is given by **x**(t) eqn (9.2.3) with $n_o = n_I$ and the number of output nodes is $n_l = m$. The activation function of the hidden nodes is typically chosen as

$$F(v(t))=\frac{1}{1+\exp(-v(t))} \tag{9.3.4}$$

or

$$F(v(t))=\frac{1-\exp(-2v(t))}{1+\exp(-2v(t))} \tag{9.3.5}$$

The output nodes usually do not contain a threshold parameter and the activation functions are linear to give

$$\hat{y}_i(t)=\sum_{j=1}^{n_{l-1}} w_{ij}^{l} x_j^{l-1}(t) \tag{9.3.6}$$

Cybenko (1989) and Funahashi (1989) have proved that the multilayered perceptron is a general function approximator and that one hidden layer networks will always be sufficient to approximate any continuous function. To simplify the notation therefore only one hidden layer networks and only single input single output (SISO) systems (m = 1, r = 1) will be considered in the present study. It is important to emphasise however that all the results are applicable to networks with several hidden layers and multi-input multi-output (MIMO) systems. The SISO restriction means that only one output neuron, that is $n_2 = 1$ is required and the index i in eqn (9.3.6) can be dropped. With these simplifications the output of the network is

$$\hat{y}(t)=\sum_{i=1}^{n_1} w_i^2 x_i^1(t) = \sum_{i=1}^{n_1} w_i^2 F\left(\sum_{j=1}^{n_0} w_{ij}^1 x_j(t) + b_i^1\right) \tag{9.3.7}$$

The weights w_i and thresholds b_j are unknown and can be represented by the parameter vector $\Theta=[\theta_1\ \theta_2 \ldots \theta_{n_\theta}]^T$. The objective of training the neural network is to determine Θ such that the discrepancy defined as

$$\varepsilon(t) = y(t) - \hat{y}(t) \tag{9.3.8}$$

called the prediction errors or residuals are as small as possible according to a defined cost function.

9.3.2 Backpropagation

The backpropagation method of training neural networks was initially introduced by Werbos (1974) and later developed by Rumelhart and McClelland (1986). Backpropagation is just a steepest decent type algorithm where the weight connection between the j'th neuron of the (k-1)'th layer and the i'th neuron of the k'th layer and the threshold of the i'th neuron of the k'th layer are respectively updated according to

$$w_{ij}^k(t) = w_{ij}^k(t-1) + \Delta w_{ij}^k(t)$$

$$b_i^k(t) = b_i^k(t-1) + \Delta b_i^k(t) \tag{9.3.9}$$

with the increment $\Delta w_{ij}^k(t)$ and $\Delta b_i^k(t)$ given by

$$\Delta w_{ij}^k(t) = \eta_w \rho_i^k(t) x_j^{k-1}(t) + \alpha_w \Delta w_{ij}^k(t-1)$$

$$\Delta b_i^k(t) = \eta_b \rho_i^k(t) + \alpha_b \Delta b_i^k(t-1) \tag{9.3.10}$$

where the subscripts w and b represent the weight and threshold respectively, α_w and α_b are momentum constants which determine the influence of past parameter changes on the current direction of movement in the parameter space, η_w and η_b represent the learning rates and $\rho_i^k(t)$ is the error signal of the i'th neuron of the k'th layer which is backpropagated in the network. Because the activation function of the output neuron is linear, the error signal at the output nodes is

$$\rho_i^l(t) = y_i(t) - \hat{y}_i(t) \tag{9.3.11}$$

and for the neurons in the hidden layer

$$\rho_i^k(t) = F'(v_i^k(t)) \sum_j \rho_j^{k+1}(t)\, w_{ji}^{k+1}(t-1) \qquad k=l-1,...,2,1 \qquad (9.3.12)$$

where F'(v) is the first derivative of F(v) with respect to v.

Similar to other steepest descent type algorithms, backpropagation is often slow to converge, it may become trapped at local minima and it can be sensitive to user selectable parameters(Billings et al 1991a).

9.3.3 Prediction Error Learning Algorithms

Estimation of parameters in nonlinear models is a widely studied topic in system identification (Ljung and Soderstrom 1983, Billings and Chen 1989a) and by adapting these ideas to the neural network case a class of learning algorithms known as prediction error methods can be derived for the multilayered perceptron. The new recursive prediction error method for neural networks was originally introduced by Chen et al (1990a, b) and Billings et al (1991a) as an alternative to backpropagation. The full recursive prediction error (RPE) algorithm is given by

$$\begin{aligned} &\Delta(t) = \alpha_m \Delta(t-1) + \alpha_g \Psi(t)\, \varepsilon(t) \\ &P(t) = [P(t-1)-P(t-1)\, \Psi(t)(\lambda I+ \Psi^T(t)P(t-1)\Psi(t))^{-1}\, \Psi^T(t)P(t-1)]/\lambda \\ &\hat{\Theta}(t) = \hat{\Theta}(t-1)+P(t)\, \Delta(t) \\ &\varepsilon(t) = y(t) - \hat{y}(t) \end{aligned} \qquad (9.3.13)$$

where α_g and α_m are the adaptive gain and momentum respectively, λ is the forgetting factor,Θ is the estimate of the parameter vector and Ψ^T represents the gradient of the one step ahead predicted output with respect to the model parameters

$$\Psi(t,\Theta)=\left[\frac{d\hat{y}(t,\Theta)}{d\Theta}\right] \qquad (9.3.14)$$

By partitioning the Hessian matrix into qxq sub-matrices, for a network with q nodes, a parallel version of the above algorithm can be derived as

$$\begin{aligned} &\Delta_i(t) = \alpha_m \Delta_i(t-1) + \alpha_g \Psi_i(t)\, \varepsilon(t) \\ & \qquad\qquad\qquad\qquad 1 \le i \le q \qquad (9.3.15) \\ &\hat{\Theta}_i(t) = \hat{\Theta}_i(t-1) + P_i(t)\, \Delta_i(t) \end{aligned}$$

where the formula for updating $P_i(t)$ is identical to that used for P(t) in (9.3.13).

A rigourous derivation of these algorithms together with implementation details, properties and comparisons with backpropagation on both simulated and real data sets is available in the literature (Chen et al 1990a, b, Billings et al 1991a, b).

Both recursive prediction error algorithms given above converge significantly faster than backpropagation at the expense of increased algorithmic complexity. Full RPE eqn (9.3.14) violates the principle of distributed computing because it is a centralized learning procedure. However, the parallel prediction error algorithm eqn (9.3.15) which is a simple extension to full RPE consists of many sub algorithms each one associated with a node in the network (Chen et al 1990b). The parallel recursive prediction error algorithm is therefore computationally much simpler than the full version and like backpropagation learning is distributed to each individual node in the network. Although parallel RPE is still algorithmically more complex than backpropagation the decrease in computational speed that this imposes per iteration is often offset by a very fast convergence rate so that overall the RPE algorithm is computationally far more efficient.

Comparisons of backpropagation and parallel RPE are available in the literature (Chen et al 1992, Billings et al 1991a) and in section 9.4.2.

9.4 Radial Basis Function Networks

Radial basis function (RBF) networks Broomhead and Lowe (1988), Chen et al (1990c,d) consist of just two layers and provide an alternative to the multilayered perceptron architecture. The hidden layer in a RBF network consists of an array of nodes and each node contains a parameter vector called a centre. The node calculates the Euclidean distance between the centre and the network input vector and the result is passed through a nonlinear function. The output layer is just a set of linear combiners. A typical RBF network is illustrated in Figure 9.3.

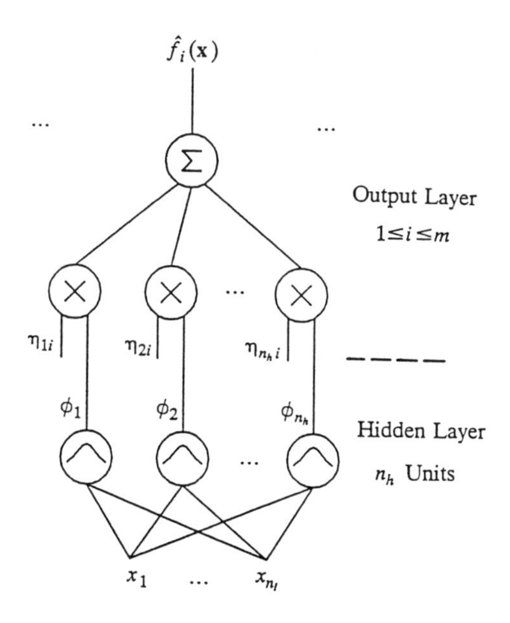

Figure 9.3 Schematic of Radial Basis Function Network

The response of the RBF network can be expressed as

$$\hat{f}_i(x) = \sum_{j=1}^{n_h} \eta_{ji} \phi_j = \sum_{j=1}^{n_h} \eta_{ji} \phi(\|\mathbf{x}-\mathbf{c}_j\|, \rho_j) \qquad 1 \le i \le m \tag{9.4.1}$$

where η_{ji} are the weights of the linear combiners, $\|.\|$ denotes the Euclidean norm, $\mathbf{c}_j$ are called RBF centres, ρ_i are positive scalars called widths, n_h is the number of nodes in the hidden layer, m=1 for SISO systems, and $\phi(\,.\,,\rho)$ are functions typically chosen as

$$\phi(z,1) = z^2 \log(z) \tag{9.4.2}$$

the thin plate spline function or

$$\phi(z,\rho) = \sqrt{z^2+\rho^2} \tag{9.4.3}$$

the multiquadratic function etc. Clearly the topology of the RBF network is very similar to the two layer perceptron with the main difference being the characteristics of the hidden nodes. Park and Sandberg (1991) have recently shown that any continuous function can be uniformly approximated to an arbitrary accuracy by a RBF network and this confirms the importance of these architectures.

9.4.1 Learning Algorithms for Radial Basis Function Networks

If a block of data is available for processing the parameters of a RBF network can be estimated using a prediction error method. This would however result in a nonlinear learning rule. Significant advantages can be gained by selecting some data points as RBF centres and developing learning rules based on linear regression methods (Broomhead and Lowe 1988). The disadvantage of this is the difficulty in selecting appropriate centres from the large number of candidate choices since an inappropriate selection will often lead to unsatisfactory results.

However, by interpreting the RBF network in terms of a NARMAX model and exploiting the extended model set idea and associated estimation algorithm of Billings and Chen (1989b) the RBF centres can be optimally positioned within a linear regression framework Chen et al (1990c,d). This means computationally cheap learning algorithms, fast adaption and avoidance of problems associated with local minima because there is only one global minimum.

The algorithm which consists of interpreting the RBF network eqn (9.4.1) in terms of the linear regression model

$$y(t)= \sum_{j=1}^{M} \phi_j(t)\, \theta_{ji} + e_i(t) \qquad 1 \leq i \leq m, \tag{9.4.4}$$

where $\phi_j(t)$ are known as regressors and θ_{ji} are the parameters to be estimated, is available in the literature (Chen et al 1990c, d).

For on-line identification applications using the RBF network a recursive algorithm which updates the centres and weights at each sample time is required. Moody and Darken (1989) showed that the n-means clustering technique of pattern classification can be used to update RBF centres. Chen et al (1991) adapted this procedure to derive a new hybrid learning algorithm for RBF networks which consists of a recursive n-means clustering sub-algorithm for adjusting the centres and a recursive least squares sub-algorithm for updating the weights. Although the new hybrid algorithm can be implemented in batch form the main advantage of the hybrid method is that it can naturally be implemented in recursive form. Space limitations preclude a full description but details are available in the literature (Chen et al 1991).

9.4.2 Backpropagation vs RPE vs the Hybrid Clustering Algorithm

Several examples which illustrate the performance of these three algorithms have been described in the literature cited above. To illustrate the typical results that are obtained consider the identification of a nonlinear liquid level system based on sampled signals of the voltage to a pump motor which feeds a set of tanks one of which is conical. This is a SISO system r=m=1. A multi-layered perceptron with one hidden layer defined by $n_I = n_y + n_u = 3+5$, $n_1 = 5$, $n_2 = 1$ giving $n_\theta = 50$ was used to model this process. The hidden node activation function of eqn (9.3.3) was selected and the initial weights and thresholds were set randomly between ± 0.3. After several trial runs it was found that $\eta_w = \eta_b = 0.01$ and $\alpha_w = \alpha_b = 0.8$ were appropriate for the backpropagation algorithm eqn's (9.3.8) and (9.3.9).

For the parallel recursive error algorithm eqn (9.3.14) a constant trace technique was used to update $P_i(t)$ and the parameters of the algorithm were set to $\alpha_g = 1.0$, $\alpha_m = 0.0$, $\lambda = 0.99$, $k_o = 60.0$ and $P_i(0) = 1000.0I$ where k_o is a constant trace technique.

The structure of the RBF network employed to identify this system was defined by $n_I = n_y + n_u = 3+5$ giving $n_h = 40$ and ϕ (.) was chosen as the thin plate spline function eqn (9.4.2). The set-up parameters for the hybrid identification algorithm were selected as $P(0) = 1000.0I$, $\lambda_o = 0.99$, $\lambda(0) = 0.95$ and $\alpha_c(0) = 0.6$, and the initial centres were randomly set. $P(t)$ is the usual matrix associated with RLS, λ_o and $\lambda(0)$ define the initial forgetting factor, and $\alpha_c(t)$ is a slowly decreasing learning rate for the RBF centres.

The evolution of the mean squared error in dB's for all three algorithms is illustrated in Figure 9.4. The results typify the performance of the three algorithms observed over several examples and clearly show that the parallel

RPE provides a significant improvement compared with backpropagation and the hybrid algorithm for RBF networks is better again.

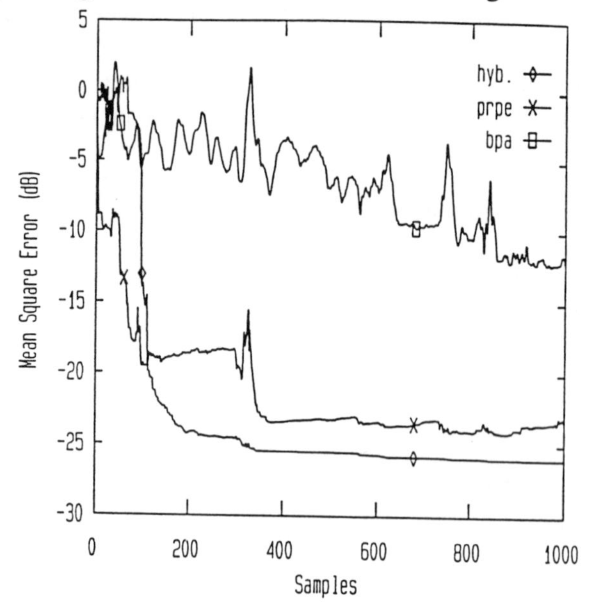

Figure 9.4 Evolution of Mean Square Error

9.5 The Functional Link or Extended Model Set Network

The main disadvantage of the MLP is that the nonlinear activation functions create problems when training the network. Nonlinear learning methods have to be applied, the learning rate can be unacceptably slow and in some applications local minima cause problems. One way to avoid nonlinear learning and hence the associated problems is to perform some nonlinear transformation or expansion of the network inputs initially and then to linearly combine the resulting terms. This is similar to the concept of RBF networks except that now there are no free parameters like the widths and centres associated with eqn (9.4.1) instead functions such as absolute value $|x_i-x_j|$, hyperbolic tanh (x_i), exponential exp $(-x_i^2)x_j$ etc etc can be employed. The potential list of functions is endless. This concept was introduced by Billings and Chen (1989b) and referred to as an extended model set (EMS). Pao (1989) discussed a similar idea and called it the functional link network (FLN). A related idea has been used by Brown and Harris (1992) to formulate the B-spline network.

Networks based on these concepts are very flexible and can be used to represent a rich class of systems with the major advantage that the learning algorithms are linear. Optimal procedures for training such networks are given in Billings and Chen (1989b).

9.6 Properties of Neural Networks

It is easy to blindly apply neural networks to data sets and to get what appeaı to be credible results. But as with other forms of identification or model fitting it is also easy to be misled and it is therefore important to be aware of the properties of the networks and the algorithms which are employed and to formulate methods which validate the results obtained.

Some properties of multilayered networks which are discussed in Billings et al (1991a,b) are briefly reviewed below.

9.6.1 Network Expansions

It is easy to see from eqn (9.3.1) to (9.3.5) that classical feed forward networks provide a static expansion of the signals assigned to the input nodes. In other words the networks do not generate lagged or dynamic components of the variables assigned as inputs to the network. This implies that if the model of a system involves an expansion of u(t-1), y(t-6) and y(t-18) say then the network model of this system is likely to be poor, irrespective of which training algorithm is employed, unless network input nodes are assigned to these variables. Input node assignment can therefore be crucial to network performance. This is not an easy problem to solve because in reality the appropriate lagged u's and y's may be distributed over a very wide range and attempts to over specify the network input nodes simply leads to increased problems of dimensionally and slow training.

9.6.2 Model Validation

Model validity tests are procedures designed to detect the inadequacy of a fitted model. Irrespective of the particular discrepancy of the model including in the neural network context incorrect input node assignment, insufficient hidden nodes, noisy data or a network that has not converged model validity tests should detect that the model is in error.

Well known tests for linear systems (Ljung and Soderstrom 1983) are inappropriate for the present application but Billings and Voon (1983,86) have shown that, for the NARMAX model, the following conditions should hold if the fitted nonlinear model is adequate

$$\begin{aligned}
&\Phi_{\varepsilon\varepsilon}(\tau) = E[\,\varepsilon(t-\tau)\,\varepsilon(t)] = \delta\,(\tau) && \forall\tau \\
&\Phi_{u\varepsilon}(\tau) = E[u(t-\tau)\,\varepsilon(t)] = 0 && \forall\tau \\
&\Phi_{u^{2'}\varepsilon}(\tau) = E[(u^2(t-\tau)-\overline{u^2}(t))\,\varepsilon(t)] = 0 && \forall\tau \qquad (9.6.1)\\
&\Phi_{u^{2'}\varepsilon^2}(\tau) = E[(u^2(t-\tau)-\overline{u^2}(t))\,\varepsilon^2(t)] = 0 && \forall\tau \\
&\Phi_{\varepsilon(\varepsilon u)}(\tau) = E[\,\varepsilon(t)\,\varepsilon(t-1-\tau\,)u(t-1-\tau\,)] = 0 && \tau\geq 0
\end{aligned}$$

Whilst it is a very difficult to prove definitively for neural networks, which are often highly nonlinear in the parameters, that the tests of eqn (9.6.1) will detect every possible model deficiency the tests have been shown to provide excellent results in practice. Normalisation to give all the tests a range of plus or minus one and approximate 95% confidence bands at $1.96/\sqrt{N}$ make the tests independent of the signal amplitudes and easy to interpret.

Consider the problem of incorrect input node assignment to illustrate the power of the model validity concept. The system defined by

$$\begin{aligned}
&\underline{y}(t) = \frac{0.4}{1+e^{-(0.3u(t-1)+0.7u(t-2)+0.5\underline{y}(t-1)+0.1)}} \\
&n(t) = e(t) + 0.6e(t-1) \qquad (9.6.2)\\
&y(t) = \underline{y}(t) + n(t)
\end{aligned}$$

was simulated with a zero mean uniformly distributed white noise input $u(t)$. Clearly this system can be represented exactly by a MLP network with just one hidden neuron and input node assignment $u(t-1),u(t-2),y(t-1)$. Initially the network was trained with the input vector incorrectly assigned as

$$\mathbf{x}(t) = [u(t-1)\;y(t-1)]^T$$

The model validity tests for this case are illustrated in Figure 9.5.

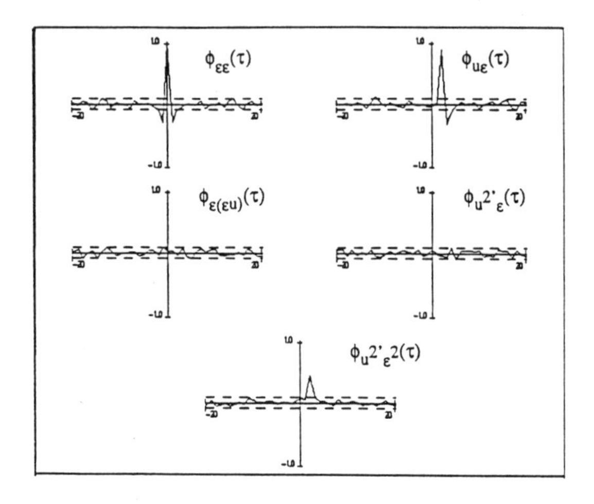

Figure 9.5 Model Validation Showing Incorrect Node Assignment

The tests $\Phi_{u\varepsilon}(\tau)$ and $\Phi_{u^{2'}\varepsilon^2}(\tau)$ are well outside the confidence bands at lag two suggesting that term of this lag has been omitted. When the network was retained with a correct input node assignment

$$\mathbf{x}(t) = [u(t-1)\ u(t-2)\ y(t-1)]^T$$

$\Phi_{u\varepsilon}(\tau)$ and $\Phi_{u^{2'}\varepsilon^2}(\tau)$ were inside the bands indicating that the input node selection was now sufficient. The autocorrelation of the residuals $\Phi_{\varepsilon\varepsilon}(\tau)$ does not satisfy eqn (9.6.1) because the simulation eqn (9.6.2) included coloured noise and $\Phi_{\varepsilon\varepsilon}(\tau)$ detects this condition which will induce bias. This is discussed in the next section.

9.6.3 Noise and Bias

Almost every single application of neural networks to the identification and modelling of nonlinear systems ignores the fact that noise will always be present if the data has been recorded from a real system. Unless the effects of the noise are understood and the noise is accommodated appropriately then incorrect or biased models will result. Bias is a well known concept in estimation theory and can often only be eliminated by fitting a noise model. If the system which is to be modelled is nonlinear then there is no reason to assume the noise will be purely linear either. The noise is likely to enter in some nonlinear way and even if it can be assumed to be additive at the system output, highly unlikely unless the system is linear, the noise on the model will in general involve cross products between the inputs and outputs Leontaritis and Billings (1985).

Bias is often difficult to detect because even if a biased network model has been obtained when this network is used to predict over the data set that was used to determine the weights and thresholds a good prediction will often be obtained. This is to be expected because the network has been trained to minimise a function of the squared prediction errors, to curve fit to the data, but it does not mean the network provides a model of the underlying mechanism.

The importance of the bias problem can be illustrated by using the example defined by

$$\underline{y}(t) = \frac{0.6}{1+e^{-(0.5u(t-1)+0.4\underline{y}(t-1)+0.1)}}$$

$$y(t) = \underline{y}(t) + n(t) \tag{9.6.3}$$

where the input u(t) was a zero mean uniformly distributed white noise sequence and n(t) represents additive noise. The form of eqn (9.6.3) has been specifically chosen to match the activation function eqn (9.3.3) so that the estimated weights can be associated with the true model parameters. Initially n(t) was set to zero and the network was trained and produced the model

$$\underline{y}(t) = \frac{0.6}{1+e^{-(0.50u(t-1)+0.398\underline{y}(t-1)+0.102)}} \tag{9.6.4}$$

A comparison of eqns (9.6.3) and (9.6.4) shows that the network is a good representation of the system.

The system was simulated again but this time with n(t) defined as coloured noise

$$u(t) = e(t) + 0.6\, e(t-1)$$

where e(t) was Gaussian white N(0,0.01). The network was retrained and produced the model

$$\underline{y}(t) = \frac{0.578}{1+e^{-(0.519u(t-1)+0.652\underline{y}(t-1)+0.091)}} \qquad (9.6.5)$$

This model is biased because the coefficient of y(t-1) is 0.652 when it should be approximately 0.4. The bias in this example only affects one parameter, in general it will alter all the weights. It is easy to detect in this example because we know the real answer.

Fortunately, the model validity tests of eqn (9.6.1), or in the time series case the tests of Billings and Tao (1991), can be used to detect these problems. The tests for the biased network of eqn (9.6.5) are illustrated in Figure 9.6.

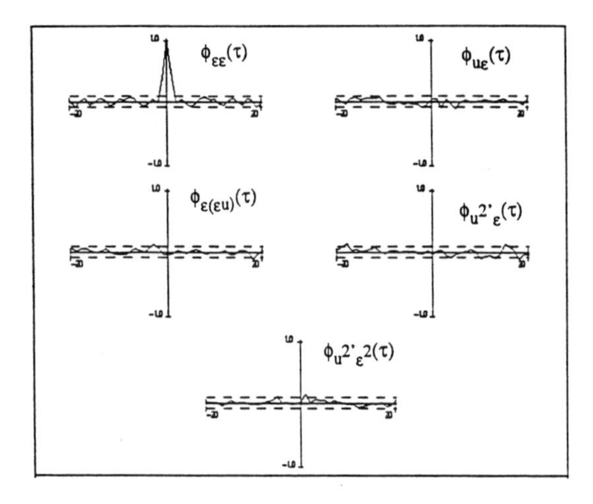

Figure 9.6 Model Validity Tests for the Network Defined by eqn (9.6.5) Showing Bias

All the correlation functions except $\Phi_{\varepsilon\varepsilon}(\tau)$ are satisfied. This indicates that the network structure is correct but that a noise model will be required to reduce the prediction errors to an unpredictable sequence and eliminate bias.

One way to avoid bias is to fit a noise model and a network architecture which can accommodate this was proposed in Billings and Jamaluddin (1991b). This consisted of augmenting the network by adding a linear mapping from lagged prediction error terms $e(t-1),..e(t-n_e)$, computed at the previous iteration, onto the usual network architecture. Alternatively a much more general solution could be obtained by simply using more input nodes in an MLP assigned to lagged prediction error terms (Chen et al 1990c).

9.6.4 Network Node Assignment

The effects of bias on the network will be influenced by the assignment of network nodes.

In general a complex nonlinear system can be represented by the NARMAX model of eqn (9.2.1). This maps past inputs and outputs into the current output according to the nonlinear expansion or in our case the network architecture. The advantage of this type of model is that it provides a very concise representation of the system. In neural network terms we would assign input nodes to represent both past inputs $u(t-1)$, $u(t-2)$... and past outputs $y(t-1),y(t-2)$.... This minimises the number of input nodes required and hence implies faster training. The disadvantage is that the parsimonious property is offset by the need to fit noise models to accommodate bias as discussed in the previous section.

The alternative is to only allow the model or the network to be a function of past inputs. So in eqn (1) all the past outputs are eliminated from the rhs to give

$$y(t) = f^{1}(u(t-1),u(t-2), \dots ,u(t-n_{u'}))+e(t) \tag{9.6.6}$$

The advantage of this is that since the noise will almost always be independent of the input even if it is coloured the estimates will be unbiased even without a noise model provided the noise is purely additive. But this property is obtained at the expense of a considerably larger network $n_{u'} >>$

n_u+n_y, increased complexity and consequently much slower training. This follows because if lagged y's are disallowed the expansion eqn (9.6.6) will involve a large, possibly infinite number, of past u's. In terms of neural networks this means considerably more input nodes in the network.

This analysis may be rather simplistic because the elimination of bias in the latter case will only be correct if the noise is additive at the output. If the system being modelled is nonlinear and if the noise occurs at internal points within the system multiplicative noise terms will be induced into the system model and biased models will be obtained unless this effect is accommodated even if the network is just an expansion in terms of past u's. When considering real systems therefore the choice of network node assignment is not clear cut. Again the model validity tests can be used to try and detect these effects.

9.6.5 Network Complexity

Overfitting by overspecifying the input node assignments and number of hidden nodes is likely to produce misleading results. It is well known in linear estimation theory that model overfitting should be avoided. The usual practice is to fit models of increasing order and to select the model of minimal complexity which just satisfies the model validity tests. These results are likely to carry over for RBF networks which are linear-in-the-parameters if the centres are known but the analysis is not so straightforward for the multilayered perceptron.

Overfitting like bias is a subtle effect. Increasing the complexity of the network is likely to improve the prediction capabilities of the network because at each stage additional terms are added which further reduce the mean squared error. But taking the extreme case, the prediction error could be reduced to zero simply by employing a look up table which associates the input at each time instant to the recorded output. It is clear that such a table or model is virtually useless, it only represents one data set. Network overfitting is just a less severe realisation of this effect Billings and Jamaluddin (1991b).

9.6.6 Metrics of Network Performance

The common measure of predictive accuracy of network performance is the one-step-ahead prediction of the system output. This was defined in eqn (9.2.4) to be

$$\hat{y}(t) = \hat{f}(y(t-1), \ldots, y(t-n_y), u(t-1), \ldots u(t-n_u)) \tag{9.6.7}$$

Inspection of eqn (9.6.7) suggests that this may not be a good metric to use because at each step the model is effectively reset by inserting the appropriate values in the rhs of (9.6.7). Any errors in the prediction are therefore reset at each step and consequently even a very poor model tends to produce reasonable one-step-ahead predictions. In reality we may wish to use the model in simulation to predict the system response to different inputs and in this situation the required output values, the y's on the rhs of eqn (9.6.7), will be unknown. It is for these reasons that we recommend that the one-step-ahead prediction should always be augmented by computing the model predicted output defined by eqn (9.2.5).

$$\hat{y}_d(t) = \hat{f}(\hat{y}_d(t-1), \ldots, \hat{y}_d(t-n_y), u(t-1), \ldots, u(t-n_u)) \tag{9.6.8}$$

Now all the output terms on the rhs are predicted not measured values, $\hat{y}_d(t)$ can be computed for any u(t) and any errors in the prediction should now become apparent because they will quickly accumulate.

9.7 System Identification

Whilst the discussions in previous sections are relevant for all applications of neural networks this section will be devoted specifically to the use of neural networks and other procedures for modelling or identifying nonlinear dynamic systems.

Neural networks provide a new and exciting alternative to existing methods of system identification. The advantages and disadvantages of these two approaches are briefly discussed below.

9.7.1 Neural Network Models

Neural networks are just another way of curve fitting to data. They have several advantages; they are conceptually simple, easy to use and have excellent approximation properties, the concept of local and parallel processing is important and this provides integrity and good fault tolerant behaviour, and they produce impressive results especially for ill defined fuzzy problems such as speech and vision. The disadvantage is that most of the current procedures ignore all the analysis of parameter estimation and they destroy the structure of the system. Some of these points will not be relevant to all applications but they certainly are for dynamic modelling. To take an extreme example to illustrate the point consider the situation where we have input/output data from a system which we wish to identify . The data was recorded from a linear system but this will not be apparent from an inspection of the data records and so an MLP network is used to model the system. The model obtained will probably predict the system output quite well but is will not reveal that the underlying system is a simple first order lag with one time constant and a gain. The neural network model destroys and does not reveal the simplicity of the underlying mechanism which produced the data. This information is distributed and encoded in the network weights and thresholds. Consequently it is difficult to analyse the fitted model to determine the range of system stability, often critical in control systems design, or the location of resonances, the sensitivity of the process output to the model time constant and gain, or to relate the model parameters to the physical components of the system.

More traditional methods of system identification overcome some of these limitations.

9.7.2 The NARMAX Methodology

Traditional methods of system identification for nonlinear systems have been based upon functional series methods (Billings 1980), block structured system models (Billings and Fakhouri 1982) and parameter estimation procedures.

Parameter estimation is of course very straightforward if the form of the model is known a priori. This information is rarely available and the NARMAX methodology (Billings 1986) is based around the philosophy that determining the model structure, or which terms to include in the model, is a critical part of the identification process. So that if the system is linear with a simple gain and time constant this should be revealed during the identification. Adopting this approach ensures the model is as simple as possible. Complex behaviour does not necessarily equate to complex models. If we do not try and determine the model structure then many different phenomena from several systems which are indeed described by the same law will appear as different laws. We will not see the simplicity and the generality of the underlying mechanism, and surely this is one of the main objectives of system identification.

The NARMAX methodology attempts to break the problem down into the following steps:-

nonlinear detection-	*is the system linear or nonlinear*
structure detection-	*which terms are in the model*
parameter estimation-	*what are the values of the unknown coefficients*
model validation-	*is the model the correct model*
prediction-	*what is the output at some future time, and*
analysis-	*what are the properties of the system.*

These components form an estimation toolkit which allows the user to build a concise mathematical description of his system that can be used as a basis for analysis and design. Both polynomial, rational and extended model set terms can be accommodated as required. The disadvantage of this approach is that the algorithms are much more complex than simple backpropagation for example, the user needs to have an understanding of parameter estimation theory and needs to interact more during the identification process.

Both the neural network and the parameter estimation approach therefore have strengths and weaknesses.

Users need to be aware of both approaches but the choice between them will often depend on the purpose of the identification. Algorithmic computing and neurocomputing therefore tend to complement each other. The former is ideal for modelling engineering systems and the like. The latter is ideal for pattern recognition, fuzzy knowledge processing, speech and vision. But there is a large area in between which will benefit from a fusion of ideas from both approaches.

9.8 Conclusions

Neural networks have opened up a whole new area of opportunity which even at this early stage of development have shown very impressive results. But we should not ignore algorithmic methods and estimation theory which offer alternative approaches and suggest solutions to many of the current problems in neural network research. A marriage of the best of both these approaches should provide a powerful toolkit for analysing an enormous range of systems and stimulate research for many years to come.

9.9 Acknowledgements

The authors gratefully acknowledge support for this work from the UK Science and Engineering Research Council.

9.10 References

Billings, S.A., 1980, Identification of nonlinear systems - a survey. *Proc. IEE, Part D*, **127**, 272-285.

Billings, S.A., 1986, An introduction to nonlinear systems analysis and identification: Chapter 10 in K. Godfrey, P. Jones (Eds): *Signal Processing for Control*; Springer Verlag, 263-294.

Billings, S.A., and Chen, S., 1989a, Identification of non-linear rational systems using a prediction-error estimation algorithm. *Int. J. Systems Sci.*, **20**, 467-494.

Billings, S.A., and Chen, S., 1989b, Extended model set, global data and threshold model identification of severely non-linear Systems. *Int. J. Control*, **50**, 1897-1923.

Billings, S.A., Fakhouri, S.Y., 1982, Identification of Systems Composed of Linear Dynamic and Static Nonlinear Elements; *Automatica*, **18**, 15-26.

Billings, S.A., Jamaluddin, H.B. and Chen, S., 1991a, A Comparison of the Backpropagation and Recursive Prediction Error Algorithms for Training Neural Networks, *Mechanical Systems and Signal Processing*, **5**, 233-255.

Billings, S.A., Jamaluddin, H.B., Chen, S., 1991b, Properties of Neural Networks with Applications to Modelling Nonlinear Dynamical Systems; *Int. J. Control*, **?**

Billings, S.A., and Leontaritis, I.J., 1981, Identification of nonlinear systems using parameter estimation techniques. *Proc. IEE Conf. Control and Its Applications*, Warwick, U.K., pp.183-187.

Billings, S.A., Tao, Q.M., 1991, Model Validity Tests for Nonlinear Signal Processing Applications, *Int. J. Control*, **54**, 157-194.

Billings, S.A. and Voon, W.S.F., 1983, Structure Detection and Model Validity Tests in the Identification of Nonlinear Systems, *Proc. IEE Pt. D*, **130**, 193-199.

Billings, S.A. and Voon, W.S.F., 1986, Correlation Based Model Validity Tests for Non-linear Models, *Int. J. Control*, **44**, No. 1, 235-244.

Broomhead, D.S., and Lowe, D., 1988, Multivariable functional interpolation and adaptive networks. *Complex Systems*, **2**, 321-355.

Brown M., Harris C. J. (1992), The B-spline neuro-controller, in '*Parallel Processing for Control*' editted by E. Rogers, Prentice-Hall

Chen, S., and Billings, S.A., 1989, Representation of non-linear systems: the NARMAX model. *Int. J. Control*, **49**, 1013-1032.

Chen, S., Billings, S.A., and Grant, P.M., 1990a, Non-linear systems identification using neural networks. *Int. J. Control*, **51**, 1191-1214.

Chen, S., Cowan, C.F.N., Billings, S.A., and Grant, P.M., 1990b, Parallel recursive prediction error algorithm for training layered neural networks. *Int. J Control*, **51**, 1215-1228.

Chen, S., Billings, S.A., Cowan, C.F.N., and Grant, P.M., 1990c, Practical identification of NARMAX models using radial basis functions. *Int. J. Control*, **52**, 1327-1350.

Chen, S., Billings, S.A., Cowan, C.F.N., and Grant, P.M., 1990d: Nonlinear Systems identification using Radial Basis Functions,' *Int. J. Systems Science*, **21**, 2513-2539.

Chen, S., Billings, S.A., and Grant, P.M., 1991, A Recursive Hybrid Algorithm for Nonlinear System Identification using Radial Basis Function Networks, *Int. J. Control*, (to appear).

Chen S., and Billings S.A., 1992, Neural Networks for Nonlinear Dynamic System Modelling and Identification, submitted to *Int. J. Control* special issue on "Intelligent Systems" 1992.

Cybenko, G., 1989, Approximations by superpositions of a sigmoidal function. *Mathematics of Control, Signals and Systems*, **2**, 303-314.

Funahashi, K., 1989, On the approximate realization of continuous mappings by neural networks. *Neural Networks*, **2**, 183-192.

Lapedes, A., and Farber, R., 1988, How neural nets works. *Neural Information Processing Systems*, edited by D.Z. Anderson (New York: American Institute of Physics), 442-456.

Leontaritis, I.J., and Billings, S.A., 1985, Input-output parametric models for non-linear systems - Part 1: deterministic non-linear systems; Part 2: stochastic non-linear systems. *Int. J. Control*, **41**, 303-344.

Ljung, L., and Soderstrom, T., 1983, *Theory and Practice of Recursive Identification*, (Cambridge: MIT Press).

Moody, J., and Darken, C., 1989, Fast-learning in networks of locally-tuned processing units. *Neural Computation*, **1**, 281-294.

Narendra, K.S., and Parthasarathy, K., 1990, Identification and control of dynamical systems using neural networks. *IEEE Trans. Neural Networks*, **1**, 4-27.

Pao, Yoh-Han, 1989, *Adaptive Pattern Recognition and Neural Networks*, (Reading: Addison-Wesley).

Park, J., and Sandberg, I.W., 1991, Universal approximation using radial-basis-function networks. *Neural Computation*, **3**, 246-257.

Rumelhart, D.E. and McClelland, J.L. (Eds.), 1986, *'Parallel Distributed Processing: Explorations in the Microstructure of Cognition,'* Vol. 1: Foundations, MIT Press.

Werbos, P.I., 1974, Beyond Regression: New Tools for Prediction and Analysis in the Behaviour Sciences: *Ph.D. thesis, Harvard University*, Cambridge MA.

Chapter 10

Neural networks: case studies

R. J. Mitchell and J. M. Bishop

10.1 Introduction

In this chapter some examples are presented of systems in which neural networks are used. This shows some of the variety of possible applications of neural networks and the applicability of different neural network techniques. The examples given also show some of the varied work done in the Cybernetics Department at the University of Reading.

10.2 Neural Network Based Vision System

On line visual inspection is an important aspect of modern day high speed quality control for production and manufacturing. If human operatives are to be replaced on production line monitoring however, the inspection system employed must exhibit a number of human-like qualities, such as flexibility and generalisation, as well as retaining many computational advantages, such as high speed, reproducability and accuracy. An important point in considering human replacement though, is total system cost, which should include not only the hardware, software and sensor system, but also commissioning time and problem solving.

A neural network learning system based on an n-tuple network (see chapter 2) is appropriate, as these are easy to use and can be implemented in hardware and so can operate very quickly. In fact the system produced, as described here, operates at video frame rate, i.e. 50 frames per second. This speed of operation is more than adequate for most production line inspection, in that up to 50 items per second can be checked as they pass a fixed point. The overall system is hardware based so that a very simple set up and start procedure is needed in order to make the system operative. This merely involves a small number of front panel switch selections with no software requirements. The overall set up is therefore low cost and simple to operate.

10.2.1 N-tuple vision system

A description of n-tuple networks is given in chapter 2, and also by Aleksander et al. (1984), but a brief summary follows. In n-tuple networks a neuron is modelled by a standard memory device, as shown in figure 10.1a), and the pattern to be analysed is presented to the address lines of the memory. If that pattern is to be remembered, a '1' is written at that address; but to see if that pattern has been learnt, the data at that address is read to see if it is a '1'. These memories can be used in a practical vision system by configuring them in the form shown in the block diagram in figure 10.1b). The system analyses data patterns consisting of r bits of information, there are m neurons (or memories) and each neuron has n inputs. Initially, '0' is stored in each location in each neuron.

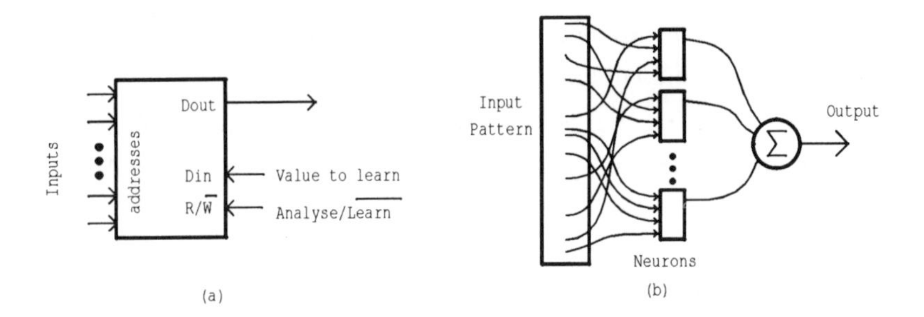

Figure 10.1 Ram Neuron and Basic n-tuple network

In the learn phase, n bits are sampled from the input pattern (these n bits are called a tuple). If the input contains analog information, then the data at any point is normally considered to be a '1' if the analog value at that point exceeds a given threshold value. This tuple is used to address the first neuron, and a '1' is stored at the specified address: effectively the neuron has 'learnt' the first tuple. Another n bits are sampled and these are 'learnt' by the second neuron. This process is repeated until the whole input pattern has been sampled. Note that it is usual to sample bits from the memory in some random manner, as this enables the system to utilise all of the spatial frequency spectrum of the image.

In the analyse phase, the same process is used to sample the input pattern so as to generate the tuples. However, instead of writing '1's into the neurons, a count is made to see how many of the neurons had 'learnt' the tuples presented to them: that is, how many neurons output a '1'. If the input pattern had already been taught to the system, the count would be 100%. If the pattern was similar, then the count would be slightly smaller.

In practice, many similar data patterns should be presented to the system, so each neuron would be able to 'learn' different tuples. Therefore the count when a pattern similar to but not identical to any member of the training set is shown to the system might still be 100% because, for example, the first tuple from the image might be the same as one from the third data pattern taught, and the second tuple might be the same as one from the seventh data pattern, etc. In general, the system is said to recognise an object if the count is greater than p%, where p is chosen suitably and is likely to depend on the application. This allows the system to deal with the inevitable noise which will affect the image.

Therefore such a system is capable of being taught and subsequently recognising data patterns. With a slight extension the system is able to be taught, recognise and discriminate between different classes of data. One set of neurons, as described above, is called a discriminator. If two sets of data are stored in one discriminator, the system will be able to recognise both sets, but not be able to discriminate between them. However, if the system has two discriminators and each data set is taught into its own discriminator, then when a data pattern is presented, a count of the number of recognised tuples is made for each discriminator, and the input pattern is most like the data class stored in the discriminator with the highest count.

An important property of the technique is that it processes all images presented to it in the same manner. Therefore it can be used in a great variety of applications.

10.2.2 A Practical Vision System

The basic aim in the design of the vision system is that it should be flexible and easy to operate. The n-tuple network method provides the flexibility, the careful system design provides the ease of operation. There are many vision systems on the market, and these often require much setting up or programming, details of which are often specific to the application. For this vision system, the intention was to allow the user to set up the system using a series of switches and light indicators. These are available on a simple front panel whose layout is shown in figure 10.2.

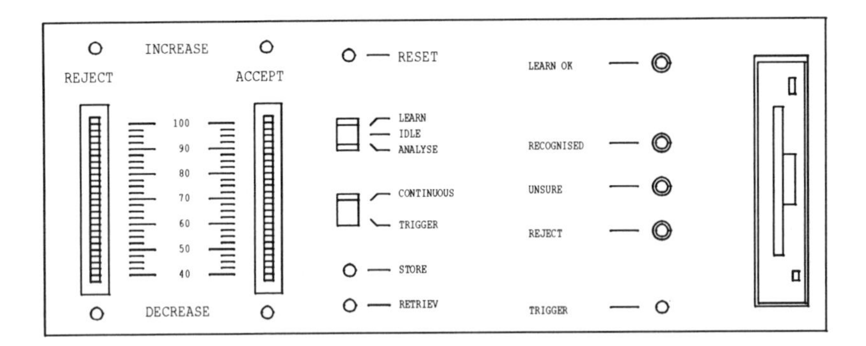

Figure 10.2 Front Panel of Vision System

Another aim of the system is low cost. Therefore in the configuration described here, there is only one discriminator: this should be enough for many applications. However, the system was designed in such a way that extra discriminators could be added easily, with a minimum of extra hardware.

The system basically operates in one of two modes: learn or analyse. In learn mode the image presented to the system is learnt, whereas in analyse mode the system indicates if the image being presented is sufficiently similar to the image or images it has been taught. Both actions are achieved by simply arranging for the object under test to be shown to the camera, the first stage of the vision system.

Each of these modes can be continuous, whereby the system processes an image at each frame time (50 frames per second), or in trigger mode, when the system only processes an image when it receives a suitable signal. These modes are selected using the switches shown in figure 10.2. An important point when learning is to teach the system a number of slightly different images, and this is easily accomplished by allowing the object being taught to slowly pass the camera along a conveyor belt; for this continuous mode is useful. However, if an object can come down the production line in various orientations, it is important to train the system by showing it the object in many orientations. This can be achieved in trigger mode, by placing the object in front of the camera, learning it by pressing the trigger button, turning the object slightly, learning the object, etc.

It is important not to saturate the memory, as then the system will recognise almost anything. Therefore the system will automatically stop learning when the discriminator is about 10% full. A light on the front panel indicates when to stop learning.

As regards analysing the object, a decision has to be made as to what is an acceptable image, and what is unacceptable, that is what percentage of recognised tuples indicates a good object. A novel feature of this vision system is that it provides an extra degree of control: it allows the user to select one level above which the system accepts the image, and a second level below which the system rejects the object. Those objects in between are marginal, and perhaps need a human to examine them to determine the appropriate action. By selecting the same level for each, the system just responds pass/fail. The levels are indicated by two LED bars shown on the front panel, and these are adjusted up or down by pressing appropriate buttons above and below the LED bars. The system response is indicated on the front panel with lights for RECOGNISED/UNSURE/REJECT, and there are corresponding signals on the back of the system which can be used to direct the object appropriately.

Although teaching an object to the system is very easy, to save time it is possible to load a previously taught object from disk and, obviously, to store such data. Loading and saving is accomplished by pressing the appropriate buttons. The data files contain both the contents of the discriminator memory and the accept and reject levels.

10.2.3 Implementation

As the system requires hardware for acquiring and processing the input video data, as well as being able to process the front panel switches and access disk drives, it was felt necessary to include in the system a micro-processor to coordinate the action. However, most of the processing is accomplished by the special hardware, so the microprocessor need not be very powerful, and its choice was largely due to the available development facilities. The n-tuple network requires various hardware modules, and the system needs to access the disk and the front panel, and so it was considered sensible to make the device in modules. A convenient method of bringing these modules together is to use a bus system, and a suitable bus is the STE bus, about which the Department has considerable expertise (Mitchell 1989). The main processor board used is DSP's ECPC which provides an IBM-PC compatible computer on one board with all the necessary signals for controlling the disk, etc.

A block diagram of the system is shown in figure 10.3. The object under study is placed in front of a standard video camera, and a monitor is used to indicate to the user that the camera is aligned correctly: this is only needed in learning mode. The output from the camera is then digitised using a module from the departmental video bus system VIDIBUS (Fletcher et al 1990), and this is then fed to the n-tuple hardware.

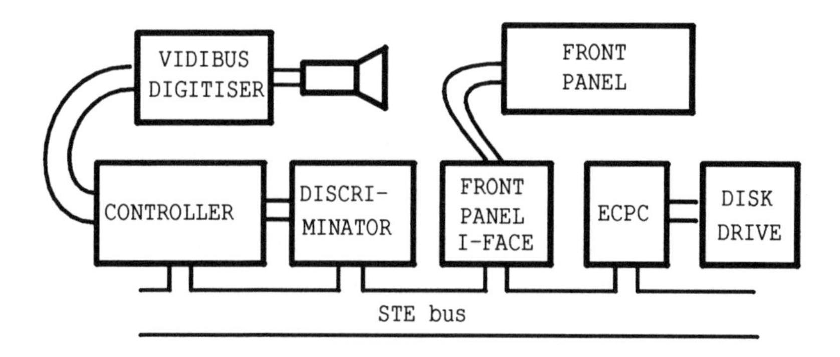

Figure 10.3 Block Diagram of Complete System

There are two parts to the n-tuple network, the controller and the discriminator. The former samples the image, forms the tuples, etc., whereas the discriminator contains the memory in which the tuples are taught and counters which indicate whether the memory is saturated (in learn mode) and the number of tuples which have been recognised (in analyse mode). The discriminator is in a separate module so as to allow a multi discriminator system to be produced relatively easily.

The ECPC is used to configure the controller, for example to tell it to process the next frame or to specify whether to learn or analyse, and to process the saturation and recognition counters. Such actions are determined by the controls on the front panel which the ECPC processes continually via a suitable interface. The last block is the disk drive, the controlling hardware/software for which is on the ECPC.

The software for the system is quite simple, as most of the processing is performed by special purpose hardware. Essentially the software monitors the front panel, configures the controller, checks the discriminator counters and processes the disk. This is all achieved by a simple program written in Turbo-C.

When developing the system, the ECPC was used as a standard PC running a compiler, etc. When the system was complete, however, EPROMs were blown containing the control program so that the system boots from the EPROM and straight away runs the program. Therefore the system does not require a disk, though it is useful for saving and reloading taught information.

10.2.4 Performance

The system has been tested on a variety of household products, including coffee jars, packets of cleaning material and hair lotion, and has been shown to operate in the expected manner. One slight problem with the system is that some care is required in setting up the lighting conditions because of the crude thresholding technique used to sample the input video signal. The Minchinton cell (Bishop, Minchinton and Mitchell 1991) can solve this problem by inserting an extra module between the digitiser and the n-tuple network. More details on the Minchinton cell are given in the section 10.3.

10.2.5 Conclusion

The system described in this section is an easy to use flexible vision system ideal for on line visual inspection. Its ease of use is due to careful system design, yet it remains flexible because of its use of neural networks. The use of standard modules, like the ECPC STE system, makes for a simple low cost solution.

10.3 A Hybrid Neural Network System

The system described above is a basic neural network used for analysing images presented to it. Although this is satisfactory for many applications, it has limitations. N-tuple networks are good at recognising patterns of a given size and position, but cannot themselves directly handle patterns of varying size and position. Recent work done in the Department as part of the BT Connectionist project (Bishop 1990) addressed this problem. This project required a system capable of determining the position of eyes on faces. This was achieved by a system which examined the input image, and tested whether the section of the image at each search point was part of an eye. N-tuple memory neurons can test the image element, but an efficient search strategy is needed also, as a simple sequential search for an eye in an image of 256*256 pixels will nominally require 65536 tests. The Stochastic Search Network (Bishop and Torr, 1991) proved useful and is described below.

The vision system described above sampled the incoming analog video signal and produced a digital value of '0' or '1' dependent on whether the analog signal exceeded a given threshold value. This rejects much information, and can cause problems when the lighting of the input image changes. The images of faces are grey level, and may occur over different lighting conditions, and so an appropriate technique is needed to handle such information. The normal method used to overcome this problem is thermometer coding (Alexander and Stonham 1979) but this requires extra memory as more neurons are needed. An alternative method is to use the Minchinton Cell (Bishop, Minchinton and Mitchell 1991): this is also described below.

10.3.1 The Stochastic Search

The Stochastic Search is a novel search technique used to obtain the best fit of a given data model within a specified search space. For this application the problem was to find an eye in an image of a face, but the same technique could be used, for example, to find the best fit of a character string in a larger string. The model and search space are defined in terms of features between which comparisons can be made; for the eye problem these features are the groups of pixels which form the tuples, but for the string problem the features could be one or more characters from the string.

The search technique uses a network of variable mapping cells to test possible locations of model features within the search space. At each iteration of the search, every cell randomly selects an associated model feature for which it defines a mapping into the search space, that is, it specifies where to look for the particular model feature in the search space. When a model feature is detected, its mapping cell becomes active and the mapping defined by that cell is maintained. If the model feature is not detected then the cell chooses a new mapping. This is accomplished by selecting another cell at random from the network, and copying its mapping if this cell is active; however, if this cell is inactive then the new mapping must be chosen randomly.

If the search conditions are favourable, a statistical equilibrium condition is soon reached whereby the distribution of mappings maintained by the cells remains constant. The most popular mapping then defines the position of the object within the search space.

Stochastic Search Networks have been shown to be very successful. For example, on average a search of 60,000 images of size 256*256 found the best match of a specified image in 24.5 iterations.

10.3.2 The Minchinton Cell

The Minchinton Cell is a general purpose preprocessor connected between the input image and the address lines of a memory neuron. It thus takes an analog input signal and produces a digital '0' or '1'. The Cell comes in various forms. In the following examples I[p] specifies a value sampled randomly at point p from the input image I:

a) $I[p] > Threshold$; this is the standard threshold function

b) $I[p_1] > I[p_2]$; this is the 'Type 0' cell

c) $I[p_1] - I[p_1+1] > I[p_2] - I[p_2+1]$; the 'Type 1' cell

The first form of cell is that used in the vision system earlier, but the so called Type 0 cell has a useful characteristic: the output of the cell is dependent on the relative amplitude of the point to the rest of the input, not the absolute amplitude. Thus if the background lighting level increased uniformally so that the grey level of all the points of the image were increased, then, providing there was no saturation, the output of the Type 0 cell would be unchanged from that at the earlier lighting levels. In practice all points of an image would not change by the same amount, but if this cell is coupled to an n-tuple network, the system is much more tolerant of changes in the background lighting levels than an n-tuple system using thresholding. The Type 1 cell, incidentally, can be used to make the system tolerant of local changes in the lighting levels.

10.3.4 The complete system

The system for determining the position of eyes in faces thus consists of a Stochastic Search Network used to suggest possible mappings into the search space, and an n-tuple network, with a Type 0 Minchinton Cell preprocessor, which had been taught a number of images of eyes.

The prototype system worked well: it was one of many systems employed as part of the Connectionist project, and within 150 iterations correctly found the position of 60% of the eyes, and thus performed as well as, or better than the other prototypes. The system also shows the ability of mixing neural network and other techniques in the solution of a problem.

10.4 Colour Recipe Prediction Using Neural Networks

One of the most important aspects of the quality control of manufacturing processes is the maintenance of colour of the product. The use of colour measurement for production and quality control is widespread in the paint, plastic and dyed textile industry but is also prevalent in many other areas including food stuffs. An industrial colour control system will typically perform two primary functions relating to the problems encountered by the manufacturer of a coloured product. First the manufacturer needs to find a means of producing a particular colour. This involves selecting a recipe of appropriate dyes or pigments which when applied at a specific concentration to the product in a particular way, will render the required colour. This process is known as recipe prediction and is traditionally carried out by trained colourists who achieve a colour match via a combination of experience and trial and error. Instrumental recipe prediction was introduced commercially in the 1960s and has become one of the most important industrial applications of colorimetry. The second function of a colour control system is the evaluation of colour difference between a batch of the coloured product and the standard on a pass/fail basis.

The first commercial computer for recipe prediction (Davidson, Hemmendinger and Landry, 1963) was an analog device known as COMIC (Colorant MIxture Computer) but all colour systems on the market today employ digital computers. A typical colour control system consists of a reflectance spectrophotometer connected to a digital computer and costs between £20,000 and £50,000. All computer recipe prediction systems developed commercially to date are based on an optical model that relates the concentrations of individual colorants to some measurable property of the colorant in use (e.g. reflectance). The model must also describe how the colorants behave when used in mixtures with each other.

The model that is almost exclusively used is the Kubelka-Munk theory (Judd and Wyszecki, 1975). It relates measured reflectance values to colorant concentrations via two terms K and S, which are the Kubelka-Munk version of the absorption and scattering coefficients of the colorant. This theory is a highly simplified version of rigorous radiative transfer theory (Chandrasekhar 1950) whereby only two fluxes of radiation are considered. Attempts have been made to introduce more complex theories by using three or more fluxes (Mudgett and Richards 1971), but the application of these more complex theories is generally not practical (Mehta and Shah 1987). The use of the exact theory of radiation transfer is not of practical interest to the coloration industry. However, for the Kubelka-Munk approximation to be valid many restrictions are assumed (Nobbs 1986).

There are many applications of the Kubelka-Munk theory in the coloration industry where these assumptions are known to be false. In particular the applications to thin layers of colorants, for example, lithographic printing inks (Westland 1988) and fluorescent dyestuffs (Ganz 1977 and McKay 1976) have generally yielded poor results.

10.4.1 Neural Networks and Recipe Prediction

The performance of the Kubelka-Munk theory in certain areas of coloration is such as to warrant investigation into potential alternative approaches. The trained colourist accumulates experience of the behaviour of the colorants and is able to extrapolate and interpolate from this experience to predict recipes for new shades without the use of Kubelka-Munk theory or any other algorithmic model. It was hoped that a suitable multi-layer perceptron network would be able to automatically learn relationships between colorants and colour, and hence learn to predict which colorants, and at which concentrations, need to be applied to a particular substrate in order to produce a specified colour. Preliminary results of this work have been published elsewhere (Bishop, Bushnell and Westland 1990, 1991).

10.4.3 Designing the Network Architecture

The Network architecture used in these experiments consisted of an input layer where cell inputs are clamped to external values (scaled CIELAB values: CIELAB is defined later), a number of hidden layers, where cell inputs are defined by the weighted activation values from the cells in the previous layer that they are connected to, and an output layer connected to the last hidden layer. This is shown diagramatically in figure 10.4.

10.4.3.1 Hand crafted architectures: combining experiential knowledge with simple heuristics

It has been shown by Funahashi (1989) that any classification task can be solved by a two layer back propagation network using a single layer of hidden units. However, if the number of hidden units exceeds the number

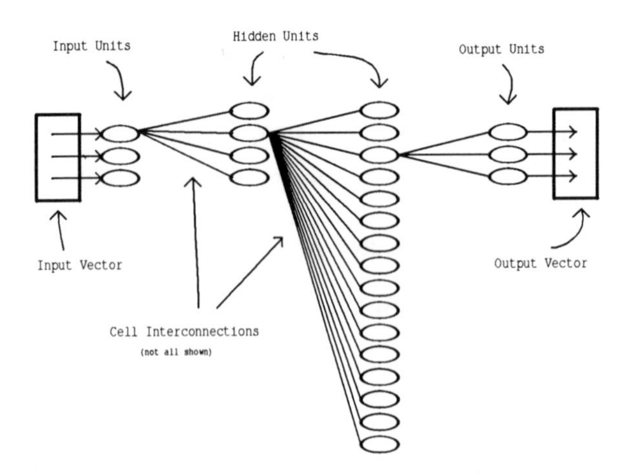

Figure 10.4 Network Architecture

of training patterns then the network may store knowledge in a very localised manner, with each hidden unit responding strongly to one training pattern. Clearly, this type of network is impractical when there are many training patterns. In addition, such a network is less able to perform useful generalistion as each unit becomes highly tuned to one training pattern. What is required is a multi-layered network which distributes knowledge throughout the net, such that the training patterns are represented by a pattern of activation across the network. Unfortunately at present there is no well defined algorithmic procedure that can be followed to generate such a network, thus the network designer either has to employ an optimisation procedure to design the network or use the rules of thumb that have evolved in place of formal heuristics.

Rumelhart (1988) describes one simple rule of thumb: networks with multiple layers and fewer units in the first layer generalise better than shallow networks with many units in each layer. Another rule is that the network should be larger than the minimum necessary to perform the task (ie. larger than that required to just learn the training data).

Even with the use of such knowledge, the eventual network design is still largely a product of inspired guesswork and experience. It may be necessary to design many such networks by hand in order to generate one useful net.

What is required is an automated method of network design which is capable of optimising network parameters such as the number of layers and the number of neurons needed in each layer.

This strategy presents some problems with generalisation. Using experience and the rules of thumb outlined above, it is possible to produce networks that can learn the training data reasonably well, (total summed squared error (TSS) of the order 10^{-3} to 10^{-4}). However, such networks often fail to generalise well for unseen data - in the case of the recipe prediction system, failing to accurately predict recipes that they have not been taught. Clearly, a network that can only predict recipes which are already known is of limited practical value!

10.4.3.2 Genetic optimisation of network architecture

Dodd (1990) describes the use of a programming strategy known as Genetic Algorithms to optimise the design of network architectures. Genetic algorithms use evolutionary techniques to perform an optimising search whose search space is defined by a binary coding of the parameters to be optimised. In the case of neural networks, the genes can represent the numbers of neurons in each layer of the network. An initial population of networks is generated randomly and tested to obtain a measure of the merit of each initial net. The test involves training each network for a short period, in relation to the usual total training time, to obtain a measure of the total error over all the training patterns. This error is then used to calculate a merit function for the net. In the case of the recipe prediction problem, the training period chosen was 20000 epochs, which is short compared to the usual training period of 200000 epochs, but long enough for the network to begin to converge if a solution can be found.

Once the initial population has been created and tested, evolutionary techniques are used to improve the population. At each generation, a parent network is chosen from the population using a weighted roulette wheel such that each net has a probability of being chosen proportional to its merit after training. The genetic string is then manipulated by one of a number of operations (also chosen on a weighted random basis) to produce a child.

The offspring is tested in the same way as the initial population to determine its merit. If the new network is better than the worst network in the population then it survives and the old network is removed from the population.

10.4.4 *Effectiveness of a Neural Network Recipe Prediction System*

10.4.4.1 Performance using hand crafted neural network architectures

The first results to suggest that neural networks might be capable of performing recipe prediction came from a set of training data synthesised by an ICS-Texicon Colour system. The colour coordinates of a selection of recipes were calculated according to standard Kubelka-Munk theory using the ICS-Texicon colour measurement system. The sets of dye concentrations and colour coordinates were then treated as real world data. About half the data obtained in this way was used to train a network whose hidden layers comprised of 8 units followed by 16 units. The input data to the network consisted of CIELAB colour co-ordinates expressed in a three dimensional cartesian format. Here, the **L** value refers to the lightness of the colour, **A** to the redness/greeness of the colour (positive **A** refers to redness and negative **A** to greeness), and **B** refers to the yellowness or blueness of the colour. The outputs from the network corresponded to the concentrations of the three dyes used in the experiment. All the numbers involved were scaled to the range 0.00 to 1.00, limited by the range of the sigmoid activation function used in the network.

The results obtained in this way were reasonable, with approximately 60% of all the predictions having a Δ value (a measure of colour difference) of less than one (0.8 is the figure normally used in making pass/fail decisions on colour samples). In most of the cases where Δ was large, it was observed that the prediction required one or more of the dyes to have zero concentration. There is an inherent difficulty when using the network to learn mappings which require some of the output units to achieve zero activation, since the sigmoid activation function needs extremely large negative inputs to achieve zero output.

To reduce this problem, all further experimentation involved scaling the output data to fit the interval 0.1 to 0.9, reducing the need to learn values at the extremities of the activation function. Experiments performed using this scaling mechanism showed a significant improvement, with approximately 80% of all predictions having Δ values of less then 0.8. This order of performance is similar to that obtained by conventional recipe prediction systems.

It is interesting to note that even with the improved scaling mechanism, the network still predicted multi-dye recipes much better than single or two dye recipes, having a failure rate of only 6.5% for three dye recipes. This situation is the reverse of that which occurs in conventional instrumental colour systems, in that the approximated Kubelka-Munk theory works better when few dyes are involved.

10.4.4.2 Performance using a genetically optimised network architecture

An optimised network was created by running the genetic algorithm program under the following conditions. The initial population consisted of ten genetic strings which were used to create the networks. These networks were each tested by running under the neural network program for 20000 cycles. The genetic algorithm then ran for 20 generations, with each generation creating 2 new offspring and removing the two worst networks from the population. Finally, the best network found by the algorithm was trained for a further 60000 cycles and compared with the best hand-crafted network trained for the same total number of cycles (80000).

Although in the initial stages of training the hand-crafted network had a smaller error than the genetically optimised one, by the end of the 80000 cycles the G.A. designed network performed significantly better than the hand-crafted one. The G.A. network had a final error of 0.000427 while the hand-crafted network had an error of 0.003931, almost 10 times worse. Unfortunetely neither network was effective enough when predicting untaught recipes (most Δ values being significantly greater than 0.8). Although networks have been able to predict synthesised recipes well (Bishop, Bushnell and Westland 1990, 1991), generalisation still remains a significant problem when dealing with real data.

10.4.5 Conclusion

Neural Networks could potentially be a very useful tool for the engineer faced with a problem that is not computationally well defined. Results using a simple multi-layer perceptron network on one such problem from the Colour Industry, have demonstrated that neural network techniques can potentially be used to solve recipe prediction problems. It has been demonstrated that the Kubelka-Munk model can be approximated without any a priori knowledge about the system. There is no reason to believe that similar neural networks cannot learn the relationship between colorant concentrations and colour coordinates for practical coloration systems, and preliminary results from ongoing research indicate that this is indeed the case (Bishop, Bushnell and Westland 1990, 1991).

In the field of Colour Recipe Prediction, the use of neural networks offers several potential advantages over the conventional approach.

i: The network can be trained on real production samples. Most dye-houses, for example, maintain historical shade data and this would be most suitable for training. The conventional Kubelka-Munk systems necessitate the preparation of special data base samples and up to ten samples per colorant is not unusual.

ii: By allowing the network to continue to learn after the initial training period, it will have the potential to adapt to changes in the production process, in a similar manner to the way that a colourist would adapt to such changes over time.

iii: The neural network approach may be able to learn the behaviour of colorants for coloration systems for which the mathematical descriptions are complex. For example, fluorescent dyes and metallic paint systems, are currently difficult to treat using standard Kubelka-Munk theory.

However the use of Neural Networks is not without problems. At present, there is no well defined set of heuristics that enable the engineer to specify an optimal network architecture and it may be difficult or impossible to learn the solution to a particular problem using an unsuitable architecture. Initial research involved the use of hand designed network architectures. Many of the designs did not learn the colour data at all, and the performance of those that did was not perfect. Recent research has involved the use of

Genetic Algorithms to optimise networks. They have several advantages and disadvantages;

i. They are very computer intensive, taking many days of CPU time on a SUN Sparc Station to converge.

ii. They can produce networks that over learn the data - that is they are able to reproduce the training data very accurately but are unable to generalise over unseen data.

iii. Networks designed by GA's are inherently very stable. That is, learning is insensitive to the initial random weights used in a given training run.

Other recent developments that have been reported in network design include the use of networks that dynamically change their topology (Hirose, Yamashita and Hijiya 1991) and the use of Gaussian noise to improve generalisation performance (Sietsma and Dow 1991). The use of both these techniques on the recipe prediction problem is ongoing.

10.5 Kohonen Network for Chinese Speech Recognition

Speech recognition is an area of research in which neural networks have been employed. In particular, Teuvo Kohonen, one of Europe's foremost contributors to neural computing, has successfully applied his Kohonen networks to this problem (Kohonen 1988). This work has been largely concerned with the understanding of Finnish speech, but this section describes briefly the work done in the Department by Ping Wu concerning Chinese speech. This is given in more detail in Wu, Warwick and Koska (1990) and Wu and Warwick (1990).

10.5.1 Review of Kohonen networks

A brief description of Kohonen networks is given here, but more details can be found in Kohonen (1984). These networks belong to the class of unsupervised learning networks, that is they learn to identify classes of data without being explicitly instructed, whereas the n-tuple and multi layer perceptron types require an external teacher. The basic structure and operation of these networks is as follows.

The network consists of a number of neurons, each neuron has a number of associated values (weights) and the input to the system is connected to each neuron. These neurons are effectively arranged in a rectangular grid, the outputs of which form a 'feature map'. These are shown in figure 10.5. Initially the weights of the neurons are given random values.

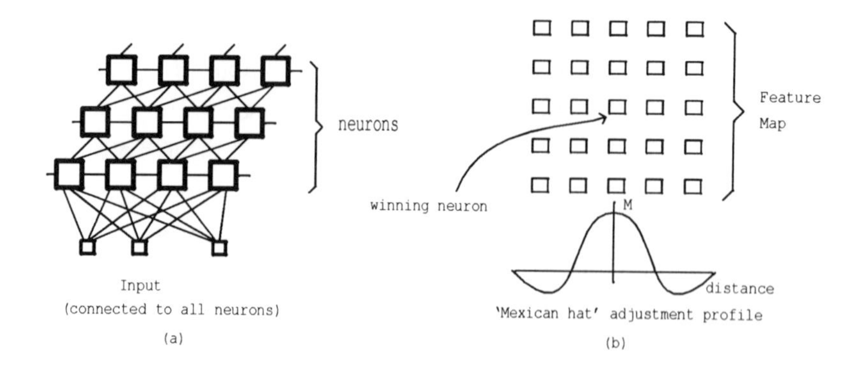

Figure 10.5 Kohonen Network

In use, the values associated with each neuron are compared with the input and the neuron which most closely resembles the input is the one which 'fires'. The weights of this neuron are then adjusted so that this neuron even more closely resembles the input. This assists the neuron to learn that pattern. However, the system must also be able to recognise similar patterns, so those neurons near the winning neuron also have their weights adjusted, but the amount by which any neuron is adjusted is proportional to the distance measured between it and the winning neuron. If the main neuron is adjusted by Δ, the other neurons are adjusted by $M*\Delta$, where M is defined by the 'Mexican hat' profile shown in figure 10.5.

As a result of the learning process, certain adjacent neurons will respond to one type of input, other neurons to other inputs, etc. In such a 'feature map' there are distinct areas of 'classes'. Thus when a pattern is input to the network, one neuron will be most active, and the system will have classified that input according to the area of the feature map in which that neuron occurs.

10.5.2 Use in speech recognition

Many forms of speech analysis categorise speech by a series of phonemes. These are the basic speech sounds, and they are concatenated to form words and ultimately sentences. Kohonen networks are used therefore to classify various inputs as to their phonemes.

If a word or sentence is then input, then various neurons will fire in a particular sequence. This appears as a trajectory in the feature map. Thus if the same word is spoken on different occasions, the system will follow approximately the same trajectory, and it is relatively easy to recognise such trajectories, and hence recognise words.

10.5.3 Chinese phoneme characteristics

Western languages concentrate on the pronunciation of complex words, so each word is formed by a number of phonemes. The Chinese language concentrates on the shape of Chinese square characters. Different shapes have different meanings, but utterance of each character takes a very simple form. There are 36 phonemes, and these can represent all of the Chinese character pronunciations. This makes Chinese a good choice for study, and has enabled an improvement to be made to the Kohonen technique.

In addition, there are only three rules determining the pronunciation of Chinese characters. The phonemes are grouped as follows:

Group 1 : b, p, m, f, d, t, n, l, g, k, h, j, q, x

Group 2 : zh, ch, sh, r, z, c, s

Group 3 : i, u, u"

Group 4 : a, o, e, ai, ei, ao, ou, an, en, ang, eng, ong

The first two are the initial consonant groups, the others are vowel sounds. The rules for pronouncing Chinese are:

A single expression is one phoneme from groups 2, 3 or 4.

A diphthong is a phoneme from group 3 followed by one from group 4.

Or, the character is one phoneme from group 1 or 2, followed by one from group 3 or group 4, or one from each group 3 and then one from group 4.

All Chinese characters obey these simple rules, so that no more than three basic phonemes are used for any character.

10.5.4 Neural network system

A Neural network to recognise Chinese speech has been simulated on a SUN computer system. The input to the system is digitised natural speech stored as phonemes. The network consists of 20 * 20 neurons. The Kohonen learning strategy can then be used to classify the phonemes.

As described in Wu and Warwick (1990), the normal feedback method used to reinforce learning (that is to adjust the winning neuron and others near it, according to the 'mexican hat' profile), can be improved upon for Chinese speech.

The mexican hat profile is symmetrical, and weights are adjusted purely according to the distance between a neuron and the winning neuron. However, as there are few rules specifying how the phonemes are grouped together, coupling of phonemes can be achieved. Thus if one phoneme fires, another particular phoneme is more likely next. Thus the feature map should be 'encouraged' to position this second phoneme near to the current one. Thus a 'bubble' shape is used in the adjustment.

Experiments have shown that the technique is successful in the recognition of Chinese speech. More details on the method can be found in the cited references.

10.6 Conclusion

The examples described in this chapter show some of the many types of neural network used in a variety of different applications. Although networks cannot be used as a solution to all projects, and that there are particular problems associated with networks to improve their performance, there is no doubt that neural networks can be a useful tool to aid the system designer when solving a particular problem.

References

Aleksander, I. and Stonham, T.J (1979); 'A Guide to Pattern Recognition using Random Access Memories', IEE Jrn.Cmp.Dig.Tch. **2**, 1, 1979.

Aleksander, I., Thomas, W. and Bowden, P (1984); 'WISARD, a radical new step forward in image recognition' Sensor Review, 120-4.

Bishop, J.M., Bushnell, M.J. & Westland, S (1990); 'Computer Recipe Prediction Using Neural Networks'. Proc. Expert Systems '90. (London).

Bishop, J.M., Bushnell, M.J. & Westland, S (1991); 'The Application of Neural Networks to Computer Recipe Prediction'. Color, **16**, 1, pp.3-9.

Bishop, J.M., Minchinton, P.R. and Mitchell, R.J (1991); 'Real time invariant grey level image processing using digital neural networks', Proc. IMechE Conf 'EuroTech Direct '91 - Computers in the Engineering Industry', Birmingham, pp 187-8.

Bishop, J.M and Torr, P (1991); 'The Stochastic Search' in 'Neural Networks for Images, Speech and Natural Language', Ed Lingard and Nightingale, Chapman and Hall.

Chandrasekhar, S (1950); 'Radiative Transfer'. Clarendon Press, Oxford.

Davidson, H.R., Hemmendinger, H. & Landry, J.L.R (1963); 'A System of Instrumental Colour Control for the Textile Industry', Journal of the Society of Dyers and Colourists, Vol.79, pp. 577.

Dodd, N (1990); 'Optimisation of Network Structure using Genetic Techniques'. Proc. INNC. '90. Paris, p 693.

Fletcher, M.J., Mitchell, R.J. and Minchinton, P.R (1990); 'VIDIBUS- A low cost modular bus system for real time video processing', Electronics and Communications Journal, **2**, 5, pp 195-201.

Funahashi, K (1989); 'On the approximate realization of continuous mappings by neural networks'. Neural Networks, **2**, 3, pp.183-192.

Ganz, E (1977); 'Problems of Fluorescence in Colorant Formulation', Colour Research and Application, **2**, pp. 81.

Hirose, Y., Yamashita, K & Hijiya, S (1991); 'Back-Propagation Algorithm Which Varies the Number of Hidden Units'. Neural Networks, **4**, 1, pp.61-66.

Judd, D.B. & Wyszecki, G (1975); 'Color in Business, Science and Industry'. 3rd ed., Wiley, New York, 1975, pp. 438-461.

Kohonen, T (1984); 'Self Organisation and Associative Memory', Springer Verlag, 1984.

Kohonen, T (1988); 'The neural phonetic typewriter', IEEE Computing Magazine, 21, 3, pp 11-22.

Mitchell, R.J (1989); 'Microcomputer systems using the STE bus', Macmillan Education, 1989.

McKay, D.B (1976); 'Practical Recipe Prediction Procedures including the use of Fluorescent Dyes'. PhD Thesis, University of Bradford (U.K).

Mehta, K.T. & Shah, H.S (1987); 'Simplified Equations to Calculate MIE-Theory Parameters for use in Many-Flux Calculation for Predicting the Reflectance of Paint Films'. Color Research and Application, Vol.12, pp. 147-153.

Mudgett, P.S. & Richards, L.W (1971); 'Multiple Scattering Calculations for Technology'. Applied Optics, **0**, pp. 1485-1502, 1971.

Nobbs, J.H (1986); 'Review of Progress in Coloration'. The Society of Dyers and Colourists, Bradford.

Rumelhart, D.E (1988); 'Parallel Distributed Processing' Plenary Session, IEEE Int. Conf. Neural Networks, San Diego, CA.

Rumelhart, D.E., Hinton, G.E. & Williams, R.J (1986); 'Learning Internal Representations by Error Propagation'. in D.E.Rumelhart, J.L.McClelland and the PDP Research Group (Eds), Parallel distributed processing: Explorations in the microstructure of cognition: Vol.1, Foundations. pp.318-362. MA: Bradford Books/MIT Press.

Sietsma, J & Dow, R.J.F (1991); 'Creating Artificial Neural Networks That Generalize', Neural Networks, **4**, 1, pp.67-79.

Westland, S (1988); 'The Optical Properties of Printing Inks'. PhD Thesis, University of Leeds, (UK).

Wu, P., Warwick, K. and Koska, M (1990); 'Neural network feature map for Chinese phonemes', Proc. Int. Conf. 'Neuronet 90', pp 357- 360, Prague.

Wu, P and Warwick, K (1990); 'A new neural coupling algorithm for speech processing', Research and Development in Expert Systems VII, Proc of Expert Systems 90, pp65-69, London.

Chapter 11

The importance of structure in neural networks

N. Dodd

11.1 Why bother with structure?

Fully connected layered networks with sufficient neurons and layers can represent any mapping. An explanation of why this is so will help to understand the relevance of structure in network design.

The first hidden layer (i.e. the first layer that isn't simply an input layer) divides up the input space with hyperplanes. Each node in this first hidden layer corresponds to a single hyperplane (defined by its weights). This is easy to visualise if we consider only two input nodes. The hyperplane in this case is a line (see below).

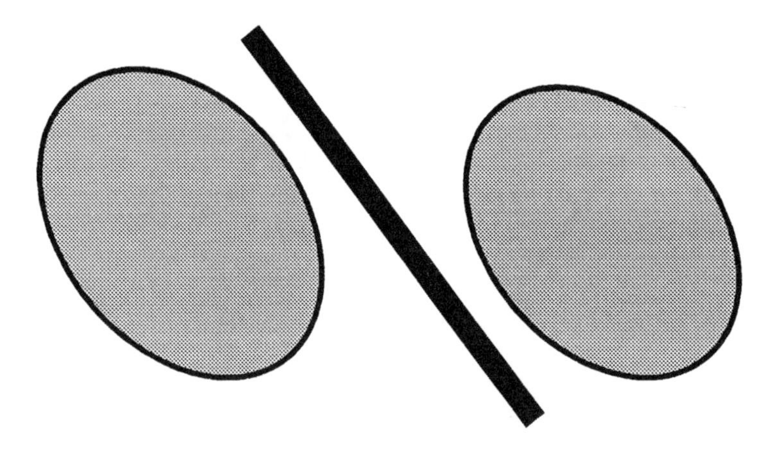

The hyperplane has the equation $\mathbf{w}.\mathbf{x} + b = 0$, where $\mathbf{w}$ is the weight vector, $\mathbf{x}$ is the input vector and b is the bias.

A number of such hyperplanes equal to the number of nodes in the first hidden layer is cast onto the input space. These hyperplanes can together form closed boundaries and if there are three layers altogether, an input layer, a hidden layer and an output layer, then it is possible for the network to separate the input space into closed regions as shown below.

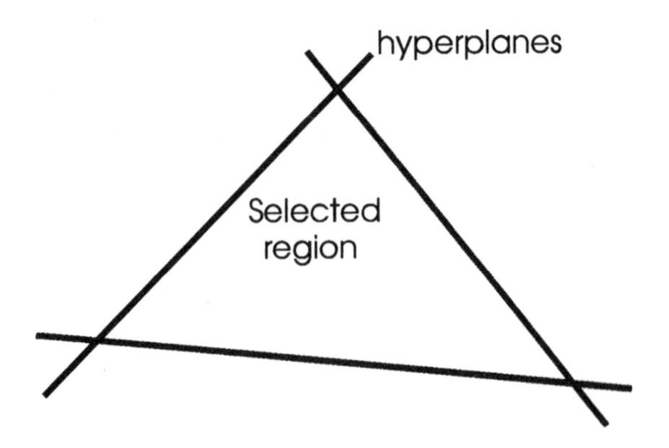

The three hyperplanes correspond to three neurons in the hidden layer and make up the surface of the region. An example of the application of this kind of network is in the Exclusive-OR problem where a two neuron hidden layer is used as described above to separate a region from the rest of input space.

Networks with yet another hidden layer can make combinations of these convex regions so that they can form regions with concavities and disconnected regions and hollow regions as shown below.

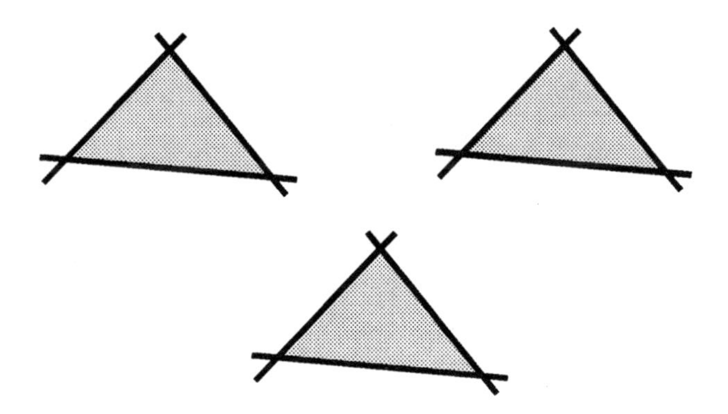

The mapping properties of the four layer fully connected network with two hidden layers are as general as could be wished for. Such a network is

On the other hand the characteristics of fast learning from any initial starting point are associated with a globally concave error surface with no local minima as shown in the next diagram. Such a weight surface means that a good solution will be found with simple error backpropagation from any initial random weight start, and is indicative of a network architecture which is well suited to the problem.

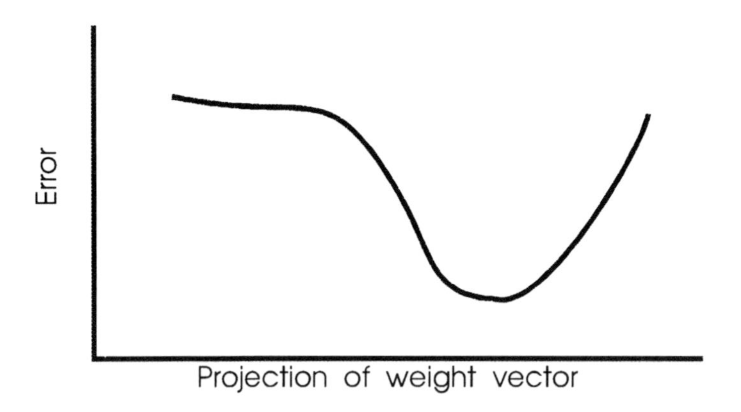

When the network architecture does not address the symmetries and invariances of the data, training is difficult since the incremental search process tends to get stuck in local minima. Learning algorithms which are less prone to get stuck have been designed, and these do sometimes work. However a network whose weight space contains many local minima also indicates that the network will not generalise well. Generalisation is fundamental to the usefulness of neural networks: a network that will correctly classify only its training set is little more than a look-up table.

The parity problem is very difficult for the fully connected layered network with logistic activation functions. However it is very easy for a network with appropriate structure. Such a network has a single neuron with a sinusoidal activation function. (Actually this network with a single sinusoidal activation function neuron can exactly map the parity mapping where the

capable of performing any mapping from input to output so long as there are sufficient nodes in each layer. So why consider any other structure of network?

There is a distinction between what can be represented by a network and what can be *learned*. The learning dynamics are closely associated with the network structure, as is the way the network will generalise or map patterns which it has not been trained with. With a network structure unsuited to the problem, learning is difficult and sometimes impossible even though we know that the network is capable of representing the mapping required. An example of this is the Parity problem where a network is required to output a 1 if there is an even number of 1s in the input and a 0 if there is an odd number. A fully connected layered network with two hidden layers is capable of representing this mapping. We can even work out the weights by hand. However some early experiments showed that it took impractically long times to train a parity network with four or more inputs, and on many occasions these training runs became stuck in local minima and would not learn at all. The learning method used in such experiments was iterative and made small changes in the weights with each training epoch. When learning has a tendency to get stuck we may imagine a graph of weight value as a function of error to have local minima as shown below

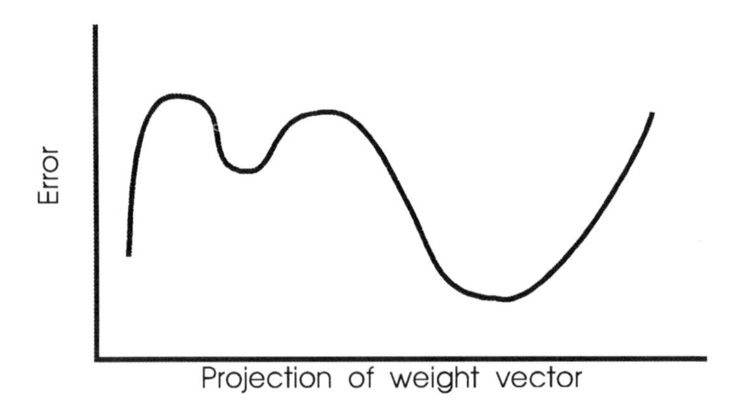

Here the lowest points on the weight surface are the most desirable. The iterative gradient descent learning process can only make downhill moves[1] in the weight space. A poorly suited network architecture will have barriers in weight space that will prevent the learning algorithm finding the good (lowest) minima. From a significant number of starting points, the learning process will become trapped in the areas of higher error in weight space.

[1]Actually, because of the finite size of the step taken in this weight space by the learning process, and because of the use of 'momentum' to speed learning, it is possible that uphill moves can occur, though infrequently.

output is either +1 or -1 (not +1 and 0 as stated above).)

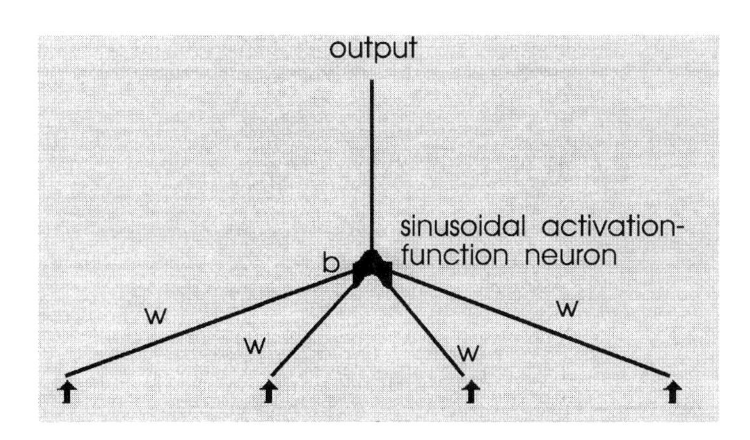

Since this network has only two independent parameters we can plot them isometrically with the corresponding sum-squared error of the neural network shown in the vertical direction:

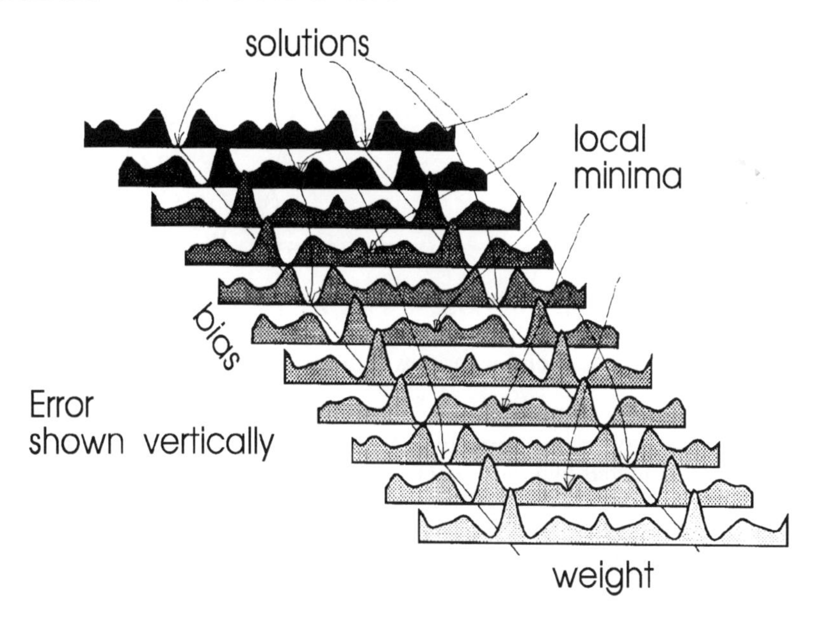

From these cross-sections, the error landscape would seem to be quite smooth. We tried starting from many hundreds of points in weight space. Simple error backpropagation found one of the solutions in 99% of cases. The diagram below shows the learning trajectory in each of these runs. The x and y axes are the weight and bias respectively, "s" denotes the start point and "x" the end point of the learning process. The dense clusters of "x" are solutions.

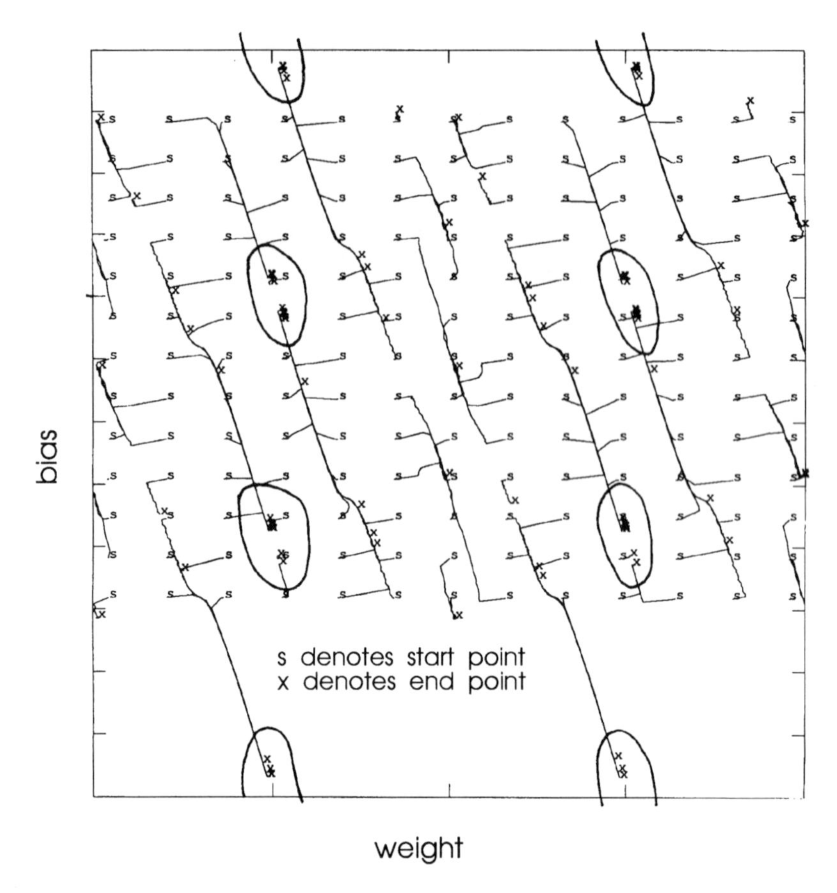

Aspects of the structure of networks which are amenable to optimisation include:

- The way the network is connected. An optimal network way not posess the fully connected layered structure beloved of many software packages.
- Activation functions of the neurons.
- Weight parameterisation. Weights may be locked, forced to have the same magnitude but opposite signs or be otherwise constrained.
- Combination of weights, w with inputs y to a neuron may be done using rules other than the dot product $x = \sum_i w_i y_i$. For instance a distance measure could be used $x = \sum_i (w_i - y_i)^2$. This gives the network very

different properties (the description of the hyperplane segmentation of the input space given at the beginning of this chapter is no longer relevant), though the chaining calculations of error-backpropagation still work.

11.2 Examples of specialised network structures

This section gives some examples of neural networks with structure which relates to their specialised problem domains.

John Shawe-Taylor [12] devised, by reference to the Perceptron Group Invariance Theorem [11], a network for identification of a graph regardless of its labeling, a problem originally posed in [3]. Interrelationships between objects are sometimes described as a graph, and the representation is used often in scene analysis. The same scene of n objects can be equally well described in $n!$ ways, simply by relabelling the objects of the scene. This makes for inefficient look-up in a table of candidate scenes. The neural network structure proposed by Shawe-Taylor maps an arbitrary labelling of the scene into a canonical form, thus enabling efficient look-up.

[4] and [2] describe the use of specially structured neural networks for classifying the state sequence of hidden Markov generators. The performance of these neural networks approaches that of the Markovian classifier (i.e. the model that generated the data when used as a classifier) and so optimally classifies the data.

[6] gives greater detail of the parity problem and its solution by a specially structured neural network than is given in this report.

[5, 7, 8] describe some comparative experiments of various ways of optimising network structure. It is shown empirically that genetic crossover is superior to other methods tried. The benchmark problem used here is underwater sonar data containing various sounds made by "dolphins".

[9] details some recent experiments using parallel computers for network optimisation. A variety of platforms are used all having Transputers as their processing units. The benchmark problem chosen for these experiments was the speech "ee-set" data from British Telecom's Connex database. This data has been used by other workers and so the results are comparable with other neural network and non-neural network experiments. The results demonstrate that an order of magnitude saving in learning time can be achieved by using optimised networks.

11.3 How do we find the appropriate structure?

For problem types where the mechanism of data production is well understood, human engineering of a suitable network structure is often possible. Where we do not have this knowledge it is desirable to make use of an automatic method for network optimisation.

An automatic search for an optimal network structure is complicated by one fundamental property of networks: the way they learn is often greatly influenced by the starting point in weight space. It is good practice to start the learning process, not with weights all of zero, but with small random values. In this way different parts of the networks are more likely to form different internal representations of the pattern space. What we want is a network architecture which is intrinsically suited to the problem class—not one which just happens to work well with some special weight start.

The influence of weight start on the fitness of a network imposes noise on the evaluation of the network. To give an experimental observation, a particular network was evaluated three times for its suitablility for a specific problem. The independent weight starts were uniformally distributed within $[-0.1, 0.1]$. The resulting error at the end of training the three networks was 0.2, 11.4 and 6.8.

We might make a biological analogy between the network specification (in terms of its unlearnable structure, i.e. connectivity, activation function etc.) and the *genotype*. The biological analogue to the network which has finished learning its training set is the *phenotype*. Thus, as in nature, the genotype is merely a recipe for a construction process which may be affected by a variety of developmental influences.

Optimisation of network architecture, or equivalently, the search of the space of networks for an optimal net for a specific problem, is made difficult if the evaluations of network fitness are noisy. We may already have a search space which has local traps, thus making gradient descent techniques ineffective, but the presence of a substantial amount of noise on top of this surface poses a severe complication as implied in the diagram below.

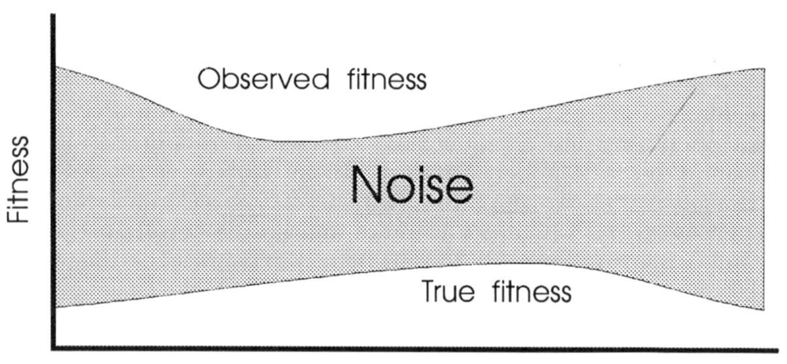

A method which will reduce the error on the estimate of the true fitness is to take repeated measurements. However each measurement of fitness requires training of a network which may require hours of computing time. We can be economical in computing resource if we only reevaluate areas in the space of networks which appear to be promising, i.e. which appear to offer relatively high fitness to within the precision of our current estimate. Such a technique was implemented, it is called *pseudo-gradient descent* and the algorithm is given below.

```
select a random genome G1
evaluate fitness, f1m, of G1
for(number of evaluations)
            {
            mutate G1, call the mutation G2
            evaluate fitness, f2, of G2
            f2m = mean(any previous fitness valuations of G2)
            if G1 = G2 then reevaluate f1m, the new mean
            if f2m > f1m then G1 = G2, f1m = f2m
            }
```

The technique of pseudo-gradient descent has its similarities to simulated annealing in that, as the algorithm progresses, the noise on the function to be minimised or maximised is reduced. With simulated annealing the noise is added intentionally, however, with the network optimisation problem, noise is added 'at source' and we must make repeated evaluations in order to see through the cloud of noise.

As an optimisation technique, genetic search [10], would seem to require excessive computational effort since an entire population of networks is maintained and each is evaluated at each generation. Comparison of genetic optimisation with the pseudo-gradient descent technique described above, and with a random search requiring the same number of evaluations, however, shows that genetic optimisation is not excessively computationally intensive. More importantly it is effective at searching this space which proves difficult for the other techniques. Analysis of genetic search shows that it evaluates each combination of parameter values contained in the gene pool at each generation. Thus an enormous number of combinations of parameter values are evaluated in parallel. This parallelism is not related to the actual implementation which may or may not be done on a parallel machine; in fact the algorthim is particularly well suited to parallel implementation which is another advantage.

The algorithm for genetic optimisation is as follows. A starting population of individuals with random genes is established. In our work there were 30 to 100 individuals in each population. Each individual network in the population was evaluated. Evaluation consists of training the network and testing it on a previously unseen test set of input patterns. The score on these patterns gave the fitness of the network. To create the next generation pairs of parents were selected with a probability proportional to their fitness. The parents were then 'mated' which consisted of laying out the parameters which describe the network in a string (see below) and selecting a random splice point. Parameters (or genes) to the left hand side of the splice point were inherited from the mother and genes from the right hand side were inherited from the father.

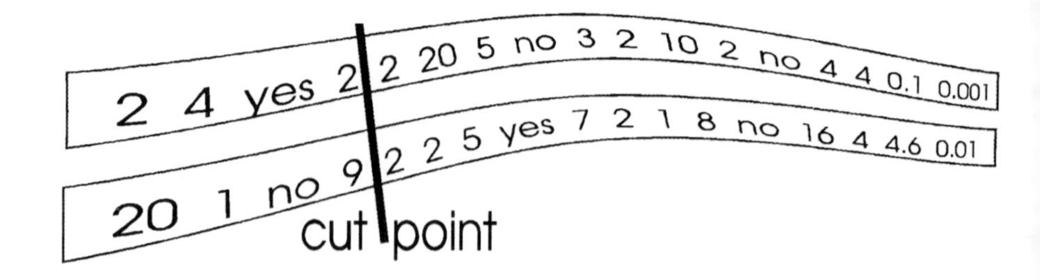

The process of evaluation followed by selection was repeated for the number of generations in the optimisation process.

In a comparative experiment we obtained the following results for the three network optimisation techniques.

	random	PGD	Crossover
optimum	10.2	19.3	17.0
re-evaluation	6.1	8.5	17.0

It can be seen that the pseudo-gradient descent algorithm finds the fittest phenotypes, whereas crossover finds the fittest genotypes. We are interested in finding the fit genotypes since we want a network description which, given any weight start, will perform well. These apparently fit individuals, however, do not perform well when re-evaluated with another set of random weight starts. The crossover technique is, by comparison, very effective at avoiding individuals which evaluate well with a special starting weight set. As expected for the crossover algorithm, the mean fitness of a re-evaluation of a generation is approximately identical to the original evaluation of fitness of the generation.

11.4 Demonstration of the advantages of an optimised network

A problem was selected which was non-trivial, used real data, and which had been used by other researchers for neural network and other recognition experiments.

Speech recognition is traditionally difficult due to the temporal and spectral variation from one utterance to another and from one speaker to another. Analytical and statistical techniques exist which are used in commercial speech recognition systems. Neural network solutions have not been successful because they have been unstructured or manually structured. The human brain constitutes an existence proof that a neural network can perform speech recognition, but it seems that the unaided human brain cannot (so far) design an *artificial* neural network which is an effective speech recogniser.

The specific recognition problem used here was to classify samples consisting of 26 frames each of 8 cepstral coefficients as one of the eight E-sounds: B, C, D, E, G, P, T or V. There were altogether 104 speakers and performance is measured on a data-set unseen in the training process.

The chosen data-set was extracted from the British Telecom Connex database. This data-set has been the subject of a large comparative study [1] in which archetypal network structures including *spread* networks and *scaly* networks were assessed.

Fitness in these experiments was defined as the number of correct guesses made by the network divided by the total number of trials. The trials were made over both the training-set and a separate unseen test-set and the final fitness determined by weighting the trials on the training-set and the trials on the test-set in a ratio of 1 : 2.

The size of the network was implicitly taken into account in the determination of fitness since each network was allowed a pre-defined fixed time for training. The smaller networks would be able to execute more training cycles in this fixed period of time than the larger networks. For two networks of equal potential but differing size it would be expected that the fitness of the smaller network would be greater since it would be allowed more training cycles.

The networks that were generated to solve the speech problem were frequently simpler than the archetypal networks devised by human beings. It turns out that a perceptron with a single layer is able to perform to within 10% of the performance of the most sophisticated network. Curiously, a network which consists of a "relay" layer of neurons which each feed from the input layer by means of connections which are all of the same weight value, and then feeds upwards to a further single perceptron layer (see below), is even more effective.

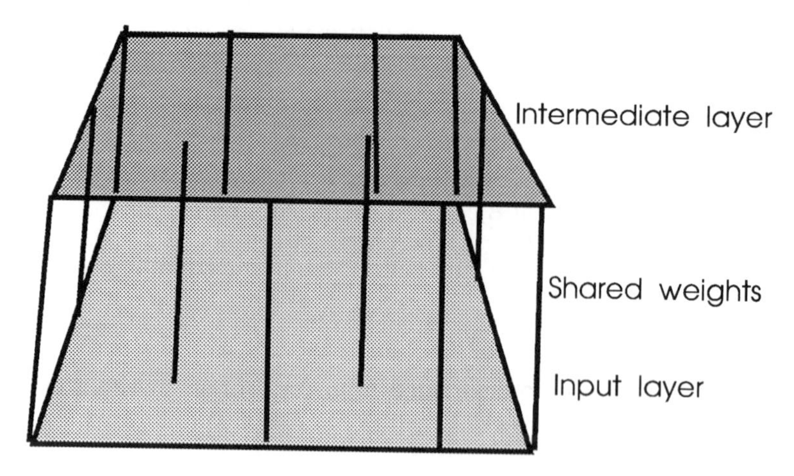

Networks which are hybrids of the types of networks used before to solve this problem were also found by the genetic optimisation process. A typical such network is shown below.

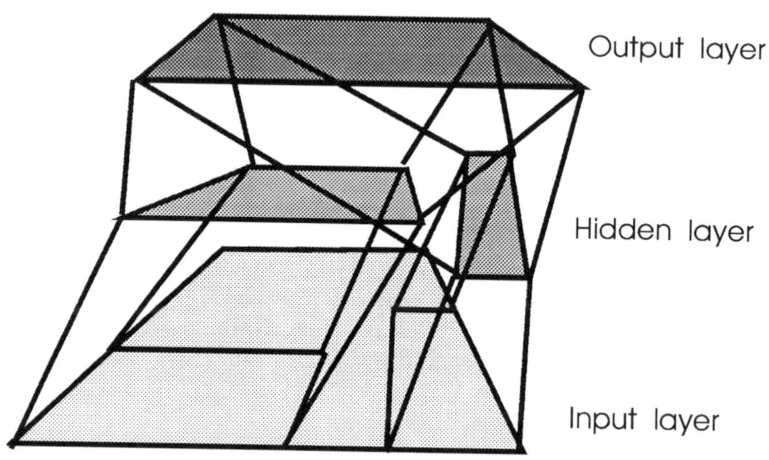

The overall performance of the genetically optimised networks was equal to the best of the networks devised manually. However the time needed to find the solution was approximately one order of magnitude less than the manually devised nets. Furthermore, the manually devised networks typically required sophisticated training algorithms, which may have been required to avoid the local minima in the weight spaces, whereas the genetically optimised networks could be trained quickly by simple gradient descent.

11.5 Conclusion

Network structure must address the symmetries and invariances of the problem for good generalisation and efficient learning. Manual network design is possible for some problem types where we understand the mechanism of data generation. For large input size and for problems where there are unknown features to the data generation mechanism an automatic method for network design must be employed. A design methodology which succeeds where others fail is genetic optimisation. This has demonstrably produced small and effective network architectures for a number of real world problems.

Bibliography

[1] Frédéric Bimbot. Speech processing and recognition using integrated neurocomputing techniques. Technical Report BRA3228, Cap Gemini Innovation, 118 rue de Tocqueville, 75017 Paris, France.

[2] N. Dodd and N.A. McCulloch. Structured neural networks for markovian processes. In *First IEE International Conference on Artificial Neural Networks*, pages 319–323, October 1989.

[3] Nigel Dodd. Graph recognition strategies. Technical Report RIPRREP/1000/50/88, Research Initiative in Pattern Recognition, 1988.

[4] Nigel Dodd. The gmlp program used as a state sequence classifier. Technical Report RIPR/1000/42/89, Research Initiative in Pattern Recognition, 1989.

[5] Nigel Dodd. Optimisation of network structure using genetic techniques. Technical Report RIPRREP/1000/63/89, Research Initiative in Pattern Recognition, 1989.

[6] Nigel Dodd. Parity implemented on an mlp with a sinusoidal activation function. Technical Report RIPRREP/1000/46/89, Research Initiative in Pattern Recognition, 1989.

[7] Nigel Dodd. Optimisation of network structure using genetic techniques. In *Proceedings of the International Joint Conference on Neural Networks*, San Diego, 1990. IEEE.

[8] Nigel Dodd. Optimisation of network structure using genetic techniques. In *Proceedings of the International Conference on Neural Networks*, Paris, 1990.

[9] Nigel Dodd, Donald Macfarlane, and Chris Marland. The generation of neural network architectures optimised for specific speech problems using genetic techniques. In *Proceedings of Eurospeech '91*, Genova, 1991.

[10] John H Holland. *Adaptation in Natural and Artificial Systems.* Ann Arbor: University of Michigan Press, 1975.

[11] M.L. Minsky and S.A. Papert. *Perceptrons.* MIT press, 1969.

[12] John Shawe-Taylor. Building symmetries into feedforward networks. In *First IEE International Conference on Artificial Neural Networks*, pages 158–162, October 1989.

Chapter 12

Pitfalls in the application of neural networks for process control

C. Hall and R. Smith

12.1 WHY USE NEURAL NETWORKS?

Like any new technique, Neural Networks must earn their keep. They must offer a solution to economically important problems that:

- is cheaper and quicker to develop than with existing techniques;
- performs at least as well, and preferably better than existing solutions;
- can be trusted.

The main attraction is in the first point; Neural Networks seem to offer a quick and easy solution to many problems that confront the control engineer. Complex systems can be modelled without

Analysis: the Neural Network can learn complex relationships - the engineer need not understand what's going on inside the black box;

Rules: because there is no understanding it cannot be codified in rules - there is no need for AI specialists to perform expensive knowledge elicitation;

Programming: Neural Network packages are sufficiently general to solve a wide range of problems without additional programming.

These promises are enough to give birth to investigative work throughout process industry; and indeed attract readers to this book. This investigative work usually takes the form of a small-scale project, designed to give the developer some experience and demonstrate the value of the technique to colleagues and management.

Many of these projects don't work or yield disappointing results. The reasons are various. As many problems stem from a lack of understanding of the subject metter as from errors in applying the technique: this chapter points out a number of errors that are fairly easy to avoid once you can recognise them.

12.2 WHERE ARE THE PITFALLS?

The pitfalls you are likely to encounter in using neural networks fall into two main groups:

- Statistical Pitfalls

These will occur when you forget to relate your knowledge (or suspicion) of statistics to the workings of the network.

- Engineering Pitfalls

These will occur when you forget that developing a system involves a number of trade-offs and a range of design options.

12.3 STATISTICAL PITFALLS

Most investigative projects will take a dataset, possibly gathered for some other purpose, and get the Neural Network to learn some feature of that dataset. We may be trying to learn a control function, or generate a process model, or characterise fault conditions. The investigator will buy one of the neural development packages, probably for his or her PC and will almost certainly build a back-propagation network with an input layer, an output layer and a hidden layer. Part of the dataset will be presented to the nodes of the input layer, and the network trained until the outputs match those desired. If the developer is competent, or just plain lucky, then the network will "converge", ie reliably learn the desired outputs, probably after several presentations of the dataset.

What are we doing? We are actually using the Neural Network to perform some curve fitting. Almost certainly we are approximating a function

$$y = f(x)$$

where x and y are input and output vectors and $\mathbf{f}$ is the relationship between them.

This curve fitting is quite a valid thing to do; it has been proved that the sigmoid functions at the outputs of the network can be combined to approximate any continuous mapping. The complexity of the resulting curve is approximately dependent on the number of hidden nodes in the network.

A caveat should be noted here; in the same way that an infinite Fourier series can approximate any continuous function, some functions may need an infinite number of hidden nodes! In practice many functions are inconvenient for a back propagation network to learn in two layers. However, we will proceed as though we have chosen a fairly well-behaved function to learn.

We will thus consider a Neural Network as a black box that performs curve fitting. The inputs are mapped via the curve to points represented by outputs (see figure 1).

We train the Network with our example data, and it converges to embody our curve. What can go wrong now?

Neural networks perform curve fitting

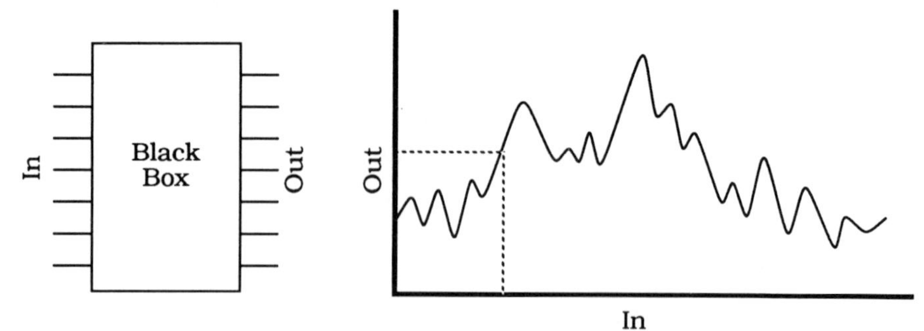

Neural network learns approximation to the function

Figure 1

We know a lot about curve fitting, but somehow we seem to forget this knowledge when we start using Neural Networks. It helps to look at how Neural Networks can:

- Overgeneralise;
- Give false confidence in extrapolations;
- Fail to learn over-complex functions;
- Fail to learn from insufficient data.

12.3.1 Overgeneralisation

The real-world data in your training set may be noisy. If you allow uncontrolled growth of the number of hidden nodes then eventually the Neural Network will learn the noise in the training set (see figure 2). When you present a new input value the output will include an error close to a similar input in the training set. It is said to be possible to calculate the "optimum" number of hidden nodes by looking at the linear independence of their weight vectors. We should, in theory, increase the number one at a time until they are just linearly dependent. The curve that is embodied in the network is now just complex enough to model the data.

Neural networks can learn the noise in your data

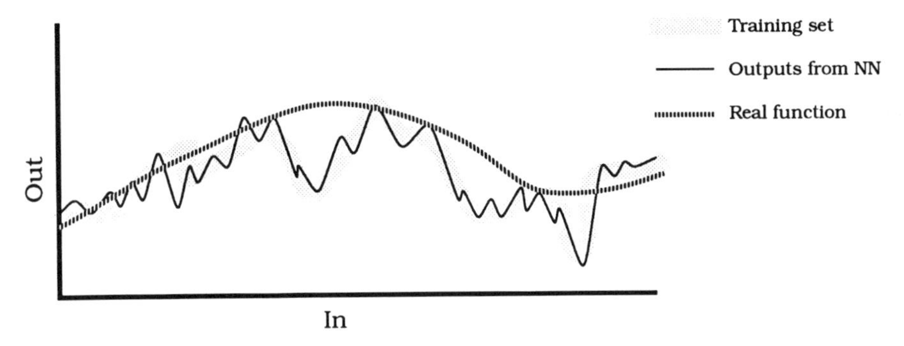

Overgeneralisation is dangerous; but you may not know it is happening

Figure 2

Unfortunately we have yet to see the Neural software package with this potentially useful feature inbuilt. We can resort to doing the painful matrix arithmetic ourselves or we can revert to heuristics.

Most beginners put too many nodes in the hidden network. We should be prepared to start with as few as two or three. One hidden node is too minimalist; you would be fitting a straight line to your data. Repeat the training cycle, increasing the number of hidden nodes one at a time. Reserve some of your data for use as a test set. Present this test data to the trained network, and compare the outputs of the network with the expected output. Plot the error against the number nodes in the hidden network (figure 3). When at a point where the error is no longer decreasing significantly, then stop.

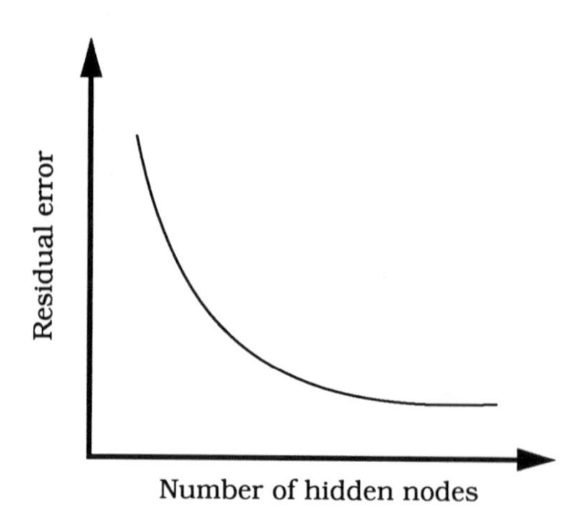

Figure 3

12.3.2 Extrapolation

Our training set may only cover part of our input space. If we ask for it to return outputs for input values which are very different to the training set, then we are either extrapolating the curve or interpolating between widely spaced "known" points (see figure 4). Because of coding it may not be obvious when we are extrapolating, and Neural Networks, like other computer programs, give a gloss of false precision to the most dubious of outputs.

Quite often developers select a test set by sampling data from the training set; eg if we took readings of our process variables every 5 minutes, we could reserve every 10th reading for the test set and use the rest to train. This means that we don't ask the network to extrapolate when testing it; but perhaps we should? It would be better to understand the limits of the network's performance now, at development rather than when it is applied to a piece of plant.

In order to define confidence regions for the network, we have to understand the training set. If that data was collected in steady-state running, it is unreasonable to expect the network to reflect what is happening in process start-up, or in moving from one feedstock to another.

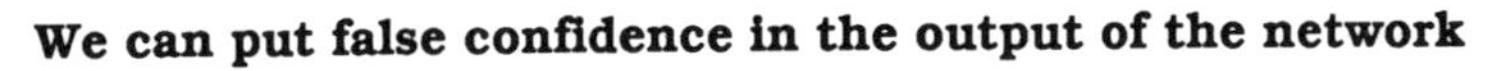

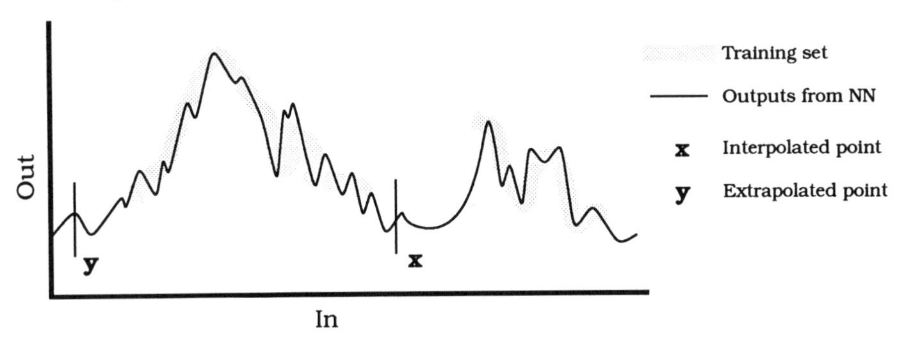

Extrapolation is dangerous; but you may not know you are doing it

Figure 4

Representing this graphically may help in relatively simple examples with few variables, but generally these are not the problems people choose to attack with Neural Networks. It is very hard to visualise a 50 dimensional input space.

12.3.4 Making Life Difficult

We may have chosen the inputs and coded them in such a way that the resulting function has gross discontinuities or gives poor separation. With better understanding of the inputs we could have chosen those which were truly important to the problem which we were trying to solve (see figure 5). Choosing and coding the features to be fed to the network is where we can apply our knowledge of the problem. It is rarely appropriate to feed a large amount of raw data to a network and expect it to abstract some order from the chaos.

How do we choose features? We could be scientific and try and do some correlation between potential input vectors and the outputs. Alternatively we probably already know a lot about the underlying physics of the relationships we are trying to model. If there are time delays in the system, then values of variables x seconds ago will be relevant. If temperature and pressure are linked, then using them both as input variables does not add any new information (and will confuse many types of network). The best feature set will be:

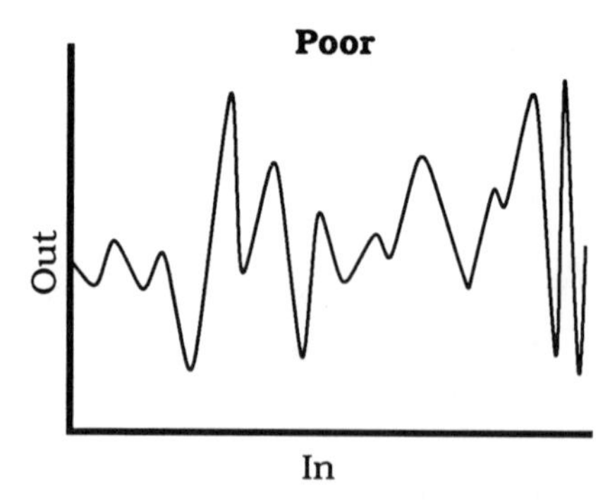

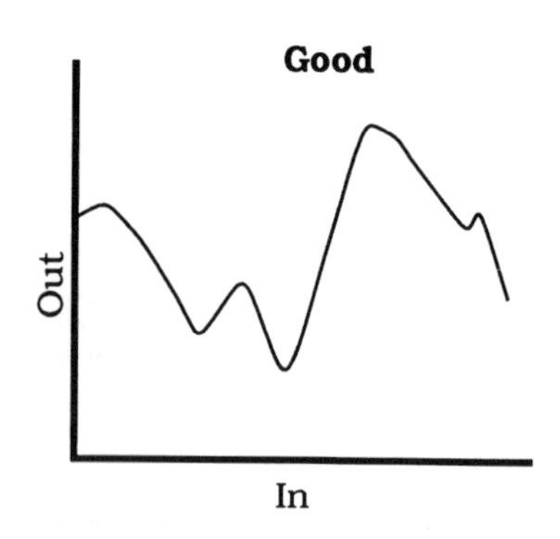

Choosing appropriate features and coding functions is important

Figure 5

- orthogonal; minimising duplication
- complete; capturing all significant influences on the outputs
- preprocessed; if you know the output depends on the square of the temperature, use T*T as an input

If you want to learn new things about the data set, then there are other, more approachable tools to use (such as BEAGLE (or Quinlan's ID3)). If you know something about the relationship that you want the Neural Network to embody, then help it on its way by your choice and coding of input features.

12.3.5 Having enough Data

John Denker of AT&T suggests that you need as many as 100 data points for every node in the hidden layer. This is a reasonable heuristic for many problems, although many researchers report quite acceptable results from relatively small training sets.

In general, the way to approach this is:

a) get as much data as possible; the more data, the better the solution;

b) design the network so that it is able to learn to perform the mapping you require to solve your particular problem.

"Generalisation" is problem specific. You can use the same, trained network and apply it to two different problems. What would be regarded as good generalisation for one problem could be no use for the other.

If you haven't got enough data points, you may be tempted to add more by superimposing an artificial "noise" function and generating (artificially) some extra data points. Unless you have specific information about the nature of noise in your data, you are simply clouding the issue. If you did know something about noise, you could build a filter; and it might be appropriate to use a Neural Network to build such a filter. If we want to consider the validity of introducing random noise into our dataset, then we have only to go back to our curve-fitting analogy (see figure 6). Introducing extra points does nothing to increase the accuracy of our curve fit, and may compromise that accuracy.

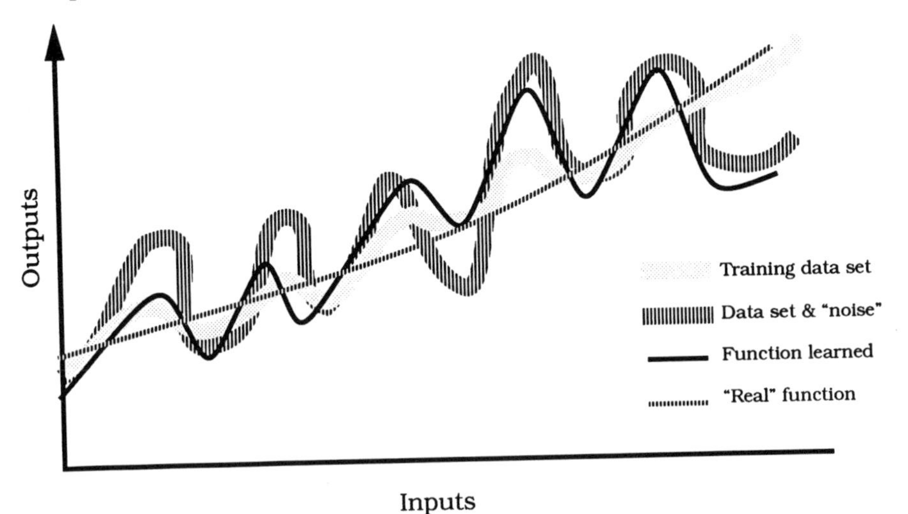

Figure 6

12.3.6 Should I have done this somehow differently?

We started this chapter by saying we were interested in economic benefits from the use of Neural Networks; benefits contrasted with other approaches. We might have used statistics, or "conventional" signal processing, or explicit numerical modelling or even knowledge-based systems, depending on our problem and our inclination. We chose Neural Networks probably because of the shrter development times that they promised. We must accept some of the disadvantages, many of which stem from the way that the Network keeps its underlying mathematics hidden, and thus makes it difficult for us to understand how it reaches its results.

12.4 ENGINEERING PITFALLS

The statistical pitfalls relate to the use of the mathematical techniques in a neural network; the engineering pitfalls concern the neural network considered as a piece of software, or as (one element in) a software system. Neural Networks can be damaging in the following ways:

- Software quality can be compromised;

- System designers can develop tunnel vision.

12.4.1 Software Quality

A perhaps unavoidable problem of using Neural Networks is that the method of developing of a working system does not sit easily with sound engineering practice.

For example, consider the conventional wisdom in software engineering. Modularity is a desirable property in a piece of software: decrease coupling between modules, increase cohesion, use it to divide development work across teams. This is confirmed by the trend towards object-orientation. Dividing the functionality of a system according to the classes (types) of things in it proves a natural and effective way of achieving a high degree of modularity. Elements of object libraries can be re-used to provide a chosen part of the functionality of an earlier system.

Neural networks are sometimes described as working on the "sub-symbolic" level and compared favourably with the "symbolic" approach to building knowledge bases, which require the contents to be articulated in detail rather than learned. Interest in the connectionist approach derives from the claim that it provides a better psychological model, especially for tasks involving a high degree of practical knowledge.

Whether you are convinced by this claim or not, you should still ask questions about a neural network considered as an engineered item and not as a surrogate human. From the point of view of software engineering, neural networks are a regressive step. Modularity is out, knowledge is distributed across all the nodes of the system and implicit in the strength of the connections and functionality emerges en bloc from a mass of training examples.

Consider the following scenario. You are purchasing a piece of software from a supplier to control a plant; it's cheap, not having taken very long to develop; your documentation consists of a list of training examples; there are no details on which functions in the source code perform a particular task; it is not possible to identify a deficiency in performance with a particular module which can be replaced; and any customisations and improvements you might want to make could effect the functioning of the whole system. How do you feel about buying this piece of software? In experimenting with neural networks you could easily end up with software systems of a similar character.

Be aware of the trade-off between on the one hand the effective development and performance of a neural network, and on the other the (software) engineering controls and standards you are accustomed to.

12.4.2 Tunnel Vision

In the light of the above criticism of neural networks, you might temporarily (re)consider alternative new techniques: fuzzy control and knowledge-based control, for example. This may save you from another pitfall: letting reports of neural network applications give you tunnel vision. The larger and more complex your application is to be, the more you should look at the possibility of combining neural networks with other techniques. You can do this in two ways: by translation or by modularisation.

It's worth remembering the strengths of representing and using knowledge in a more explicit form. It can provide a brief explanation for an action. It is more convenient for describing abstract theories or general policies which haven't developed incrementally from examples, but are hypotheses or regulations. It is public and easier to criticise, and the criticism can involve, not the accumulation of negative instances, but the production of a single counter-example sufficient to undermine all earlier confirmation.

12.4.3 Translation

You can use the learning capability of the neural network to produce information which is translated into another, more tractable, representation.

Fuzzy reasoning can be viewed as an acceptable compromise between the explicitness required of an engineered system and the implicit nature of much practical knowledge. (Note, though, that fuzzy reasoning is only implicit relative to an explicit numerical standard).

A bridge between neural networks and fuzzy reasoning is the "neuro-fuzzy" approach. For example, a neural network can be used to develop the set membership functions or calculate the weighting to be attached to clauses in fuzzy if-then rules.

There are two points to make about the translation approach. Firstly, you need to ensure that you can perform the translation: in writing the translator or altering the structure of the network to make the translation easier you may lose the gains in development time that you made in using a neural network. Secondly, you have to decide whether to allow the translated version to be revised in its translated form with development continuing using the neural network: in that case you need to able to map the revised translated version back into some network state.

12.4.4 Modularisation

You may be able to divide a system into modules capable of being implemented by separate methods. It is not necessary to turn everything into a neural network just because it can be done "in principle". Valuable development time will be lost if you have to force a neural network representation upon every piece of knowledge. The attractiveness of the neural network approach for lower level tasks can blind one to aspects of a problem where explicit knowledge is available, more appropriate and easier to encode. The architecture of an already opaque network may be rendered even more complicated by attempts to develop neural networks as a single knowledge representation formalism.

In mixed systems neural networks can be given a restricted role. For example, they can be used for recognising significant patterns in process readings, while a knowledge base of documented procedures for diagnosing problems might be encoded as rules or object methods. By limiting the tasks the neural networks perform you can assuage any worries about using a "black box" as a major component in your system; you may find it easier to document the development of a neural net module; you might even find it to be a re-usable component.

There are some concerns of the knowledge-based approach which are probably not best served by translation into a neural network regime:

- meta-knowledge; and
- explanation.

A knowledge-base may contain "meta-knowledge", that is, knowledge about knowledge. This can deal with the range of validity of an item of knowledge or decide when certain reasoning processes are done. In this way some parts of the system are intended to refer to the system itself as opposed to some external process. Used wisely, this technique can make a system more controllable and transparent. It is better suited to modelling of a person's knowledge about other ("symbolic") knowledge (for example, knowledge of rules and regulations) rather than knowledge of how to perform a task. Such meta-knowledge can be conveniently encoded using a knowledge-based approach: for example, with one rule referring to another rule or group of rules.

You may want to introduce an explanation facility as part of a system's HCI. Some conclusions or recommended actions of a system can be accepted without explanation; eg in pattern recognition where no person can articulate satisfactorily the reason for performance which we acknowledge to be effective. In cases where explanation is required, this is more naturally presented as a series of statements; these are easier to produce when you have explicit knowledge representation.

12.5 SUMMARY

Many people do build successful applications of Neural Networks; the point of outlining some pitfalls is so that they can be recognised and avoided, not to deter experimentation.

The advice we would reiterate is:

understand the data: and make sure you have enough of it. Do some preliminary analysis of the features and guess which are the most important:- give these more prominence in a coding scheme;

understand why you are building a system: in particular don't give false precision to the results. Understand (from knowledge of the training set) in what regions you can have confidence in the Neural Network, and where the wet finger is a better approximation;

use other people's experience: this is accessible in many ways. Conference and published papers rarely tell you what the developers did wrong before they finally come to a solution that worked. Practically-oriented clubs (such as LINNET) aim to tell the whole story rather than just the good points.

use your own experience: don't throw away what you have already. Neural networks can contribute to your final system without forming the whole of it. See how they relate to your existing knowledge of fuzzy reasoning, knowledge-based systems and software engineering techniques. Remember to ask for development and maintenance costs when you're told that your preferred approach can be translated into a neural network. Don't rush to buy special-purpose hardware.

If you think you can use an analytical technique, or statistics, or an explicit programming style to do what you need to do then that is preferable. The theory behind the technique's operation is understood and documented and you know what amount of confidence you can place in the results.

However, analysis may be too difficult, or too expensive, or not fast enough. In some cases a Neural Network will yield acceptable results for a very modest outlay.

Index